I0481860

Electrical Power Distribution

Dr. Hidaia Mahmood Mohamed Alassouli

Hidaia_alassouli@hotmail.com

Introduction

This book includes my lecture notes for electrical power distribution book. The fundamentals of electrical power distribution are applied to various distribution system layouts and the function of common distribution system substations and equipment. The book introduces the design procedures and protection methods for power distribution systems of consumer installations. Circuit simulation and practical laboratories are utilised to reinforce **concepts.**

The book is divided to different learning outcomes

- CLO 1- Discuss the fundamental concepts related to electrical distribution systems.
- CLO 2- Explain the role of distribution substations and related equipment.
- CLO 3- Outline standard methods for power distribution to consumer installations.
- CLO 4- Apply short-circuit and over-load protection principles for electrical installations

a) CLO1- Discuss the fundamental concepts related to electrical distribution systems.

- Explain the role of the distribution system in a power system, common distribution system layouts, and common voltages, voltage drops and regulation levels from transmission to distribution.
- Discuss demand, power quality issues, calculate factors affecting design, and interpret the load curve profile for load demand.
- Explain how tariff is calculated and charged consumers

b) CLO2- Explain the role of distribution substations and related equipment.

- Explain the function of the distribution substation in view of distribution system layout
- Explain the use of transmission, grid, primary and distribution substations a power system.
- Explain the use of various types of bus-bar configurations in distribution substations.
- Discuss the use of cabling, transformers, circuit breakers, switches, reclosers, and sectionalisers in a distribution system.

c) CLO3- Outline standard methods for power distribution to consumer installations.

- Discuss commonly used methods for low voltage power supply systems (TN, TN-C, TN-C-S and TT).
- Discuss the main features of a one-line, electrical installation diagram and related symbols.
- Discuss electrical color codes and factors affecting cable installations.
- Design an electrical feeder by (1) selecting the design current, (2) selecting the overload current protection, (3) determining the applicable correction factors, (4) selecting the current-carrying capacity of cable and cable sizing, and (5) calculating the allowable voltage drop in feeder

d) CLO4- Apply short-circuit and over-load protection principles for electrical installations.

- Explain the meaning of overload and over-current and methods of protection
- Discuss the nature of electric shock, need for earthing, earth loop impedance, and principle of protective multiple earthing.
- Explain the principles of fuse/MCB selection in relation to feeder protection under overload and short circuit fault conditions.
- Explain the operation of earth leakage circuit breakers (ELCB) and residual current device (RCD).

LO1

Discuss the fundamental concepts related to electrical distribution systems.

LO1-1

Fundamental Concepts Related to Electrical Distribution Systems

Power systems are comprised of 3 basic electrical subsystems.

- Generation subsystem
- Transmission subsystem
- Distribution subsystem

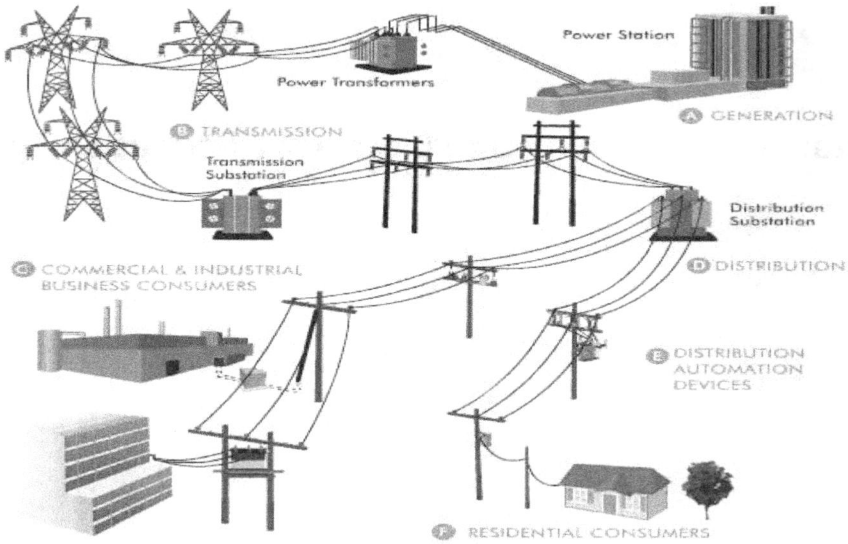

Role of the Distribution System

- The function of the electric power distribution system in a building or an installation site is to receive power at one or more supply points and to deliver it to the individual lamps, motors and all other electrically operated devices.
- The importance of the distribution system to the function of a building makes it almost imperative that the best system be designed and installed.

Role of the Distribution System

- The best distribution system is one that will, cost-effectively and safely, supply adequate electric service to both present and future probable loads.
- The distribution sub-station receives and manages voltage to a value suitable for local distribution.
- About 40% of power system investment is in the distribution system equipment (40% in generation, 20% in transmission).
- *In general*, distribution system is that part of power system which distributes power to the consumers for utilization.

Classification of Distribution Systems

(1) *Nature of current.* According to nature of current, distribution system may be classified as

(*a*) d.c. distribution system

(*b*) a.c. distribution system. Now-a-days, a.c. system is universally adopted for distribution of electric power as it is simpler and more economical than direct current method.

Classification of Distribution Systems

(2) ***Type of construction.*** According to type of construction, distribution system may be classified as

(*a*) overhead system

(*b*) underground system.

- The overhead system is generally employed for distribution as it is 5 to 10 times cheaper than the equivalent underground system.
- In general, the underground system is used at places where overhead construction is impracticable or prohibited by the local laws.

Classification of Distribution Systems

(3) ***Scheme of connection.***

- According to scheme of connection, the distribution system may be classified as

  (*a*) radial system

  (*b*) ring main system

  (*c*) inter-connected system.

- Each scheme has its own advantages and disadvantages and those are discussed later one.

Distribution System

- In general, the distribution system is the electrical system between the sub-station fed by the transmission system and the consumers meters.
- It generally consists of ***feeders, distributors*** and the ***service mains***

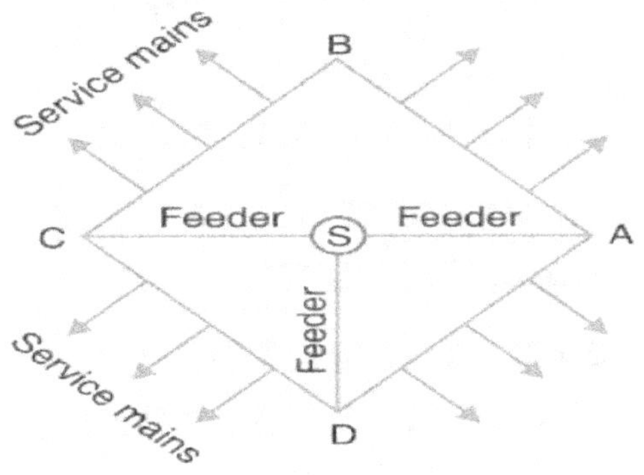

Fig1: Single line diagram of a typical low tension distribution system

Distribution System

(1) Feeders

- A feeder is a conductor which connects the sub-station (or localized generating station) to the area where power is to be distributed.
- Generally, no tappings are taken from the feeder so that current in it remains the same throughout.
- The main consideration in the design of a feeder is the *current carrying capacity*.

Distribution System

(2) Distributor

- A distributor is a conductor from which tappings are taken for supply to the consumers.
- In Fig. 1, *AB*, *BC*, *CD* and *DA* are the distributors.
- The current through a distributor is not constant because tappings are taken at various places along its length.
- While designing a distributor, voltage drop along its length is the main consideration since the statutory limit of voltage variations is ± 6% of rated value at the consumers' terminals.

Distribution System

(3) Service mains

A service mains is generally a small cable which connects the distributor to the consumers' terminals.

A.C. Distribution

- Now-a-days electrical energy is generated, transmitted and distributed in the form of alternating current.
- One important reason for the widespread use of alternating current in preference to direct current is the fact that alternating voltage can be conveniently changed in magnitude by means of a transformer.
- Transformer has made it possible to transmit a.c. power at high voltage and utilize it at a safe potential. High transmission and distribution voltages have greatly reduced the current in the conductors and the resulting line losses.
- The a.c. distribution system is the electrical system between the step down substation fed by the transmission system and the consumers' meters.

A.C. Distribution

- The a.c. distribution system is classified into

(*1*) Primary distribution system

(*2*) Secondary distribution system.

A.C. Distribution

Primary distribution system.

- It is that part of a.c. distribution system which operates at voltages somewhat higher than general utilisation and handles large blocks of electrical energy than the average low-voltage consumer uses.
- The voltage used for primary distribution depends upon the amount of power to be conveyed and the distance of the substation required to be fed.
- The most commonly used primary distribution voltages are 11 kV, 6·6kV and 3·3 kV.
- Due to economic considerations, primary distribution is carried out by 3-phase, 3-wire system.

A.C. Distribution

Primary distribution system.

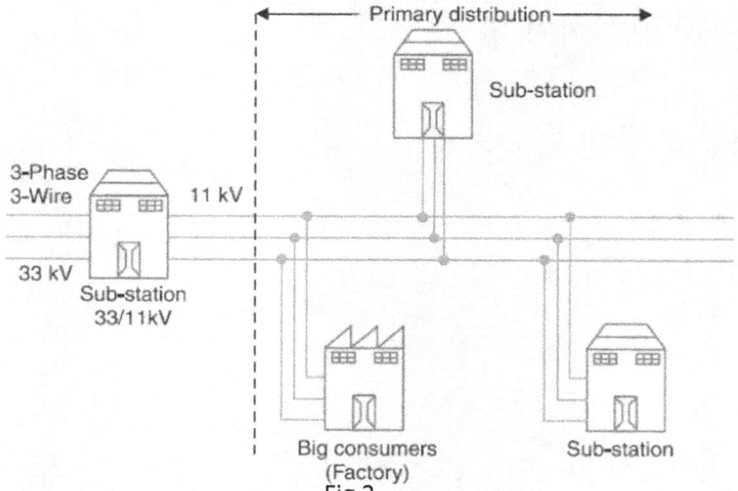

Fig 2

D.C. Distribution

2-wire distribution system

- As the name implies, this system of distribution consists of two wires.
- One is the outgoing or positive wire and the other is the return or negative wire. The loads such as lamps, motors etc. are connected in parallel between the two wires as shown in Fig. 4.
- This system is never used for transmission purposes due to low efficiency but may be employed for distribution of d.c. power.

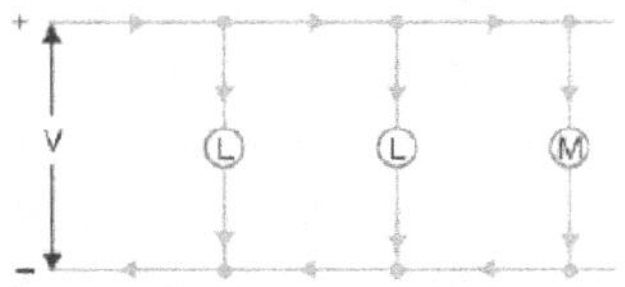

Fig 4

D.C. Distribution

3-wire distribution system

- It consists of two outers and a middle or neutral wire which is earthed at the substation.
- The voltage between the outers is twice the voltage between either outer and neutral wire as shown in Fig. 5.
- The principal advantage of this system is that it makes available two voltages at the consumer terminals.
- Loads requiring high voltage (*e.g.*, motors) are connected across the outers, whereas lamps etc. requiring less voltage are connected between either outer and the neutral.

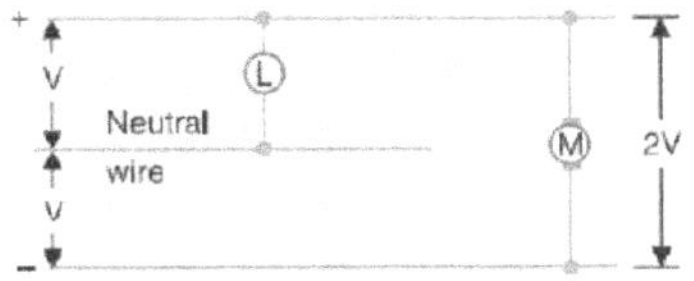

Fig 5

Voltage Classifications

- ANSI (American National Standards Institute) and IEEE® (Institute of Electrical and Electronics Engineers) standards define various voltage classifications for single-phase and three-phase systems.
- The terminology used divides voltage classes into:

❖ Low voltage

❖ Medium voltage

❖ High voltage

❖ Extra-high voltage

❖ Ultra-high voltage

Voltage Class	Nominal System Voltage	
	Three-Wire	Four-Wire
Low voltage	240/120 240 480 600	208Y/120 240/120 480Y/277 —
Medium voltage	2400 4160 4800 6900 13,200 13,800 23,000 34,500 46,000 69,000	4160Y/2400 8320Y/4800 12000Y/6930 12470Y/7200 13200Y/7620 13800Y/7970 20780Y/12000 22860Y/13200 24940Y/14400 34500Y/19920
High voltage	115,000 138,000 161,000 230,000	— — — —
Extra-high voltage	345,000 500,000 765,000	— — —
Ultra-high voltage	1,100,000	—

Voltage Classifications

- Transmission lines can carry higher voltages than the plant generates – up to 765,000 V - so a transformer steps up the voltage for the transmission lines at the transmission substation.
- Higher voltages mean less current for the same amount of power, so wires can be smaller and cost less, but higher voltages need more insulation, which is more costly and limits the voltages carried by transmission lines.

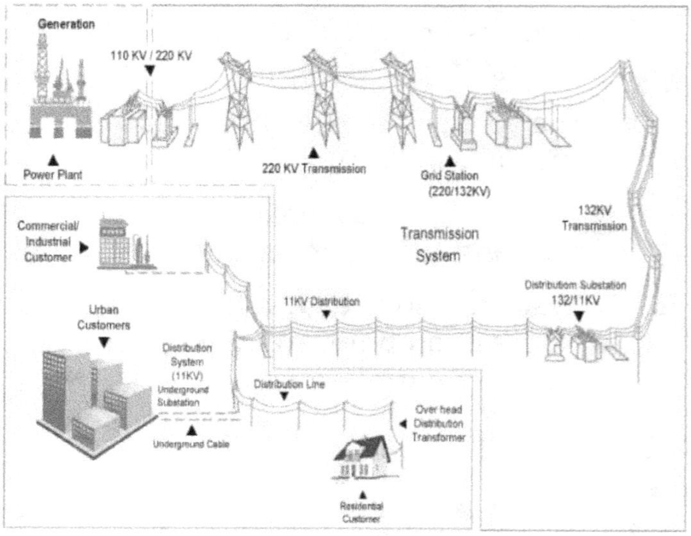

Fig 6

Voltage Classifications

- Electricity travels the transmission lines to a distribution substation, at which point it is stepped down.
- The electricity then travels on distribution lines atop wooden or concrete poles, or is run underground through special conduits.
- The electricity is stepped down again when it reaches a structure to between 120 – 240 volts.
- A cable composed typically of three wires – two live and one ground – carries the power to the structure through a meter box.

Common Distribution Layouts

- All distribution of electrical energy is done by constant voltage system. In practice, the following distribution circuits are generally used :

(1) Radial System

- In this system, separate feeders radiate from a single substation and feed the distributors at one end only.
- Fig. 7 shows a single line diagram of a radial system for d.c. distribution where a feeder *OC* supplies a distributor *AB* at point *A*. Obviouslv, the distributor is fed at one end only *i.e.*, point *A* is this case.

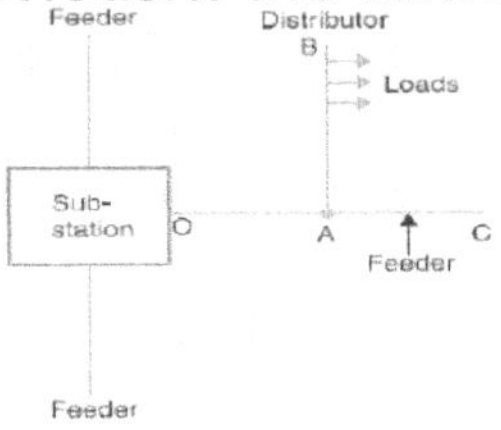

Fig 7

Common Distribution Layouts

- Fig. 8 shows a single line diagram of radial system for a.c. distribution.
- The radial system is employed only when power is generated at low voltage and the substation is located at the centre of the load.

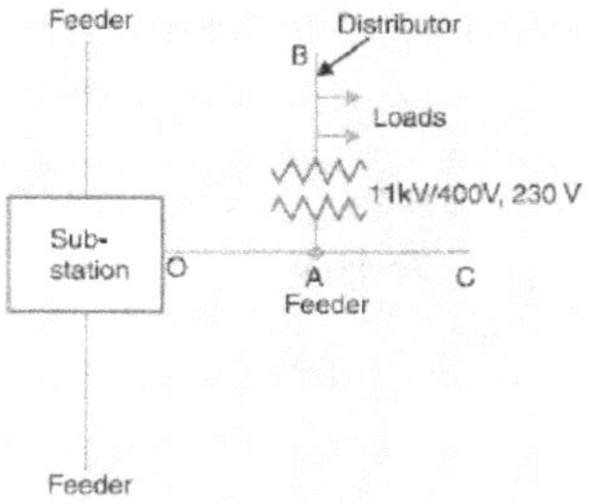

Fig 8

Common Distribution Layouts

- This is the simplest distribution circuit and has the lowest initial cost. However, it suffers from the following ***drawbacks*** :

(*a*) The end of the distributor nearest to the feeding point will be heavily loaded.

(*b*) The consumers are dependent on a single feeder and single distributor. Therefore, any fault on the feeder or distributor cuts off supply to the consumers who are on the side of the fault away from the substation.

(*c*) The consumers at the distant end of the distributor would be subjected to serious voltage fluctuations when the load on the distributor changes.

- Due to these limitations, this system is used for short distances only.
- The main ***advantages*** are:

(a) Low cost .

(b) Simple planning

Common Distribution Layouts

(2) Ring Main System

- In this system, the primaries of distribution transformers form a loop.
- The loop circuit starts from the substation bus-bars, makes a loop through the area to be served, and returns to the substation.
- Fig. 9 shows the single line diagram of ring main system for a.c. distribution where substation supplies to the closed feeder LMNOPQRS.
- The distributors are tapped from different points *M*, *O* and *Q* of the feeder through distribution transformers.

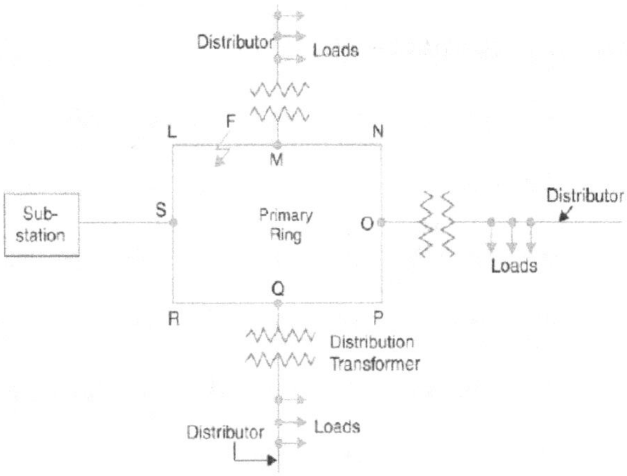

Fig 9

Common Distribution Layouts

- The ring main system has the following ***advantages***:

(*a*) There are less voltage fluctuations at consumer's terminals.

(*b*) The system is very reliable as each distributor is fed *via* two feeders. In the event of fault on any section of the feeder, the continuity of supply is maintained.

For example, suppose that fault occurs at any point *F* of section SLM of the feeder. Then section SLM of the feeder can be isolated for repairs and at the same time continuity of supply is maintained to all the consumers *via* the feeder *SRQPONM*.

Common Distribution Layouts

(3) Interconnected System.

- When the feeder ring is energised by two or more than two generating stations or substations, it is called inter-connected system.
- Fig. 10 shows the single line diagram of interconnected system where the closed feeder ring *ABCD* is supplied by two substations *S*1 and *S*2 at points *D* and *C* respectively.
- Distributors are connected to points *O*, *P*, *Q* and *R* of the feeder ring through distribution transformers.

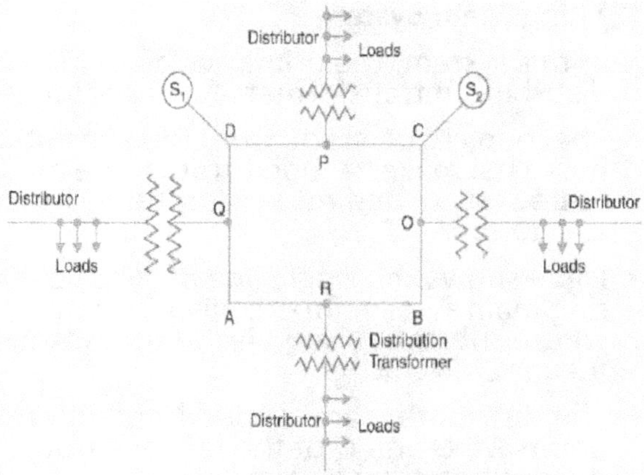

Fig 10

Common Distribution Layouts

- The interconnected system has the following ***advantages*** :

(*a*) It increases the service reliability.

(*b*) Any area fed from one generating station during peak load hours can be fed from the other generating station. This increases efficiency of the system.

Overhead Versus Underground System

- The distribution system can be overhead or underground.
- ***Overhead lines*** are generally mounted on wooden, concrete or steel poles which are arranged to carry distribution transformers in addition to the conductors.
- The ***underground*** system uses conduits, cables and manholes under the surface of streets and sidewalks.
- The choice between overhead and underground system depends upon a number of widely differing factors.

Overhead Versus Underground System

(*i*) *Public safety.* The underground system is more safe than overhead system because all distribution wiring is placed underground and there are little chances of any hazard.

(*ii*) *Initial cost.* The underground system is more expensive due to the high cost of trenching, conduits, cables, manholes and other special equipment. The initial cost of an underground system may be five to ten times than that of an overhead system.

(*iii*) *Flexibility.* The overhead system is much more flexible than the underground system. In the latter case, manholes, duct lines etc., are permanently placed once installed and the load expansion can only be met by laying new lines. However, on an overhead system, poles, wires, transformers etc., can be easily shifted to meet the changes in load conditions.

Overhead Versus Underground System

(*iv*) *Faults.* The chances of faults in underground system are very rare as the cables are laid underground and are generally provided with better insulation.

(*v*) *Appearance.* The general appearance of an underground system is better as all the distribution lines are invisible. This factor is exerting considerable public pressure on electric supply companies to switch over to underground system.

(*vi*) *Fault location and repairs.* In general, there are little chances of faults in an underground system. However, if a fault does occur, it is difficult to locate and repair on this system. On an overhead system, the conductors are visible and easily accessible so that fault locations and repairs can be easily made.

Overhead Versus Underground System

(*vii*) *Current carrying capacity and voltage drop.* An overhead distribution conductor has a considerably higher current carrying capacity than an underground cable conductor of the same material and cross-section. On the other hand, underground cable conductor has much lower inductive reactance than that of an overhead conductor because of closer spacing of conductors.

(*viii*) *Useful life.* The useful life of underground system is much longer than that of an overhead system. An overhead system may have a useful life of 25 years, whereas an underground system may have a useful life of more than 50 years.

(*ix*) *Maintenance cost.* The maintenance cost of underground system is very low as compared with that of overhead system because of less chances of faults and service interruptions from wind, ice, lightning as well as from traffic hazards.

(*x*) *Interference with communication circuits.* An overhead system causes electromagnetic interference with the telephone lines. The power line currents are superimposed on speech currents, resulting in the potential of the communication channel being raised to an undesirable level. However, there is no such interference with the underground system.

Overhead Versus Underground System

- Comparative economics (*i.e.*, annual cost of operation) is the most powerful factor influencing the choice between underground and overhead system.
- The greater capital cost of underground system prohibits its use for distribution.
- But sometimes non-economic factors (*e.g.,* general appearance, public safety etc.) exert considerable influence on choosing underground system.
- In general, overhead system is adopted for distribution and the use of underground system is made only where overhead construction is impracticable or prohibited by local laws.

Power system quality

The Power system Quality expresses to which degree a practical supply system resembles ideal supply.

If power quality is good, then any loads run satisfactory and efficiently. Installation running cost and carbon footprint will be minimal.

Basic Power System Problems

1 Transient

An undesirable event that is undesirable and momentary in mature.

1.1 Impulsive Transient

Sudden non power frequency change in steady state conditions of voltage or current (i.e. lightening is impulsive transient).

1.2 Oscillatory Transient

Consists of voltage and currents whose instantaneous value changes polarity rapidly. It is defined by its spectral content, duration and magnitude. These are generally associated with ferroresonance and transformer energization or capacitor energization.

Principe of Overvoltage Protection

The main sources of the transient over voltage protection are capacitor switching, magnification of capacitor switching transients and lightening.

The fundamental principles of overvoltage protection of load equipment are:

- Limit the voltage across sensitive insulation.
- Divert the surge current a way from the load.
- Block the surge current from entering the load.
- Bond ground references together at the equipment.
- Reduce or prevent surge current from flowing between grounds.
- Create a low pass filter using blocking or limiting principles.

The surge arresters and transient voltage surge supressors are widely used. Their main function is to limit the voltage that can appear between two points in the circuit.

Short Duration Voltage Variation

1 Interruption

Occurs when supply voltage or load currents decrease to less than 0.1 pu for period of time not greater than 1 min. It results of the power system faults and equipment failure.

2 Sags

A decrease to between 0.1 to 0.9 in rms. voltage or current for duration from 0.5 cycle to 1 min. They are caused by energization of heavy load or starting of large motors i.e. typical voltage sag can be associated with SLG fault on another feeder or substation.

3 Swell

Increase between 1.1 and 1.8 p.u. in rms. voltage or current of power frequency for duration from 0.5 cycles to 1 min. It is associated with system faults. One way when the voltage sag can occur for temporary voltage rise on unfaulted phase during SLG and also can be caused by switching off large load or energising large capacitor.

Protection against Voltage Sags and Interruption

As we entertain solutions at higher levels, solution is more costly. The solution close to the load is cheaper.

- The least cost solution is often for the end user to specify to the supplier that the machine is able to ride through sags of a designated duration. Many suppliers can provide the necessary capability if it is specified at the time quotations are requested.
- At the next higher level, it may be possible to apply an uninterruptible power supply (UPS) system or some other type of power conditioning to the machine control.
- At higher level 3 some sort of backup power supply with the capability to support the load for brief period is required.
- Higher level 4 represents alterations to the utility power system to significantly reduce the number of sags and interruptions.

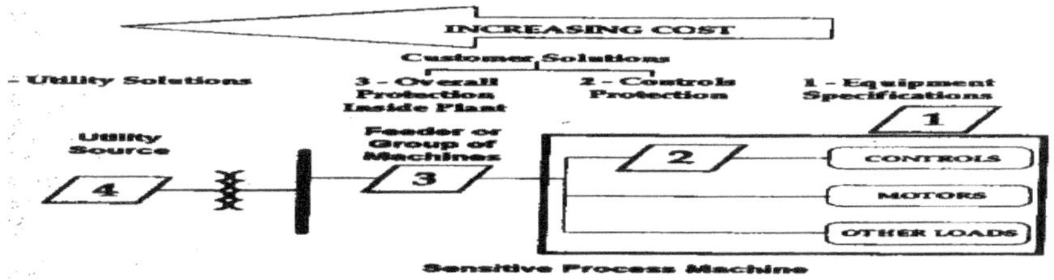

Long Duration Voltage Variation

1 Overvoltages

Increase in rms. ac voltage greater than 110% at the power frequency for longer than 1 min. It results from load switching (e.g. switching off large load or energising large capacitor). It results because the system is too week for voltage regulation or voltage control is inadequate.

2 Undervoltages

Decrease in voltage less than 90% at the power frequency for duration longer than 1 min. A load switching on or a capacitor bank switching off can cause an undervoltage until the voltage equipment on system can bring voltage back to within tolerances.

3 Sustained Interruption

When supply voltage has been zero for longer time than 1 min, the long duration voltage variation is considered as sustained interruption

Protection against Long Duration Voltage Variations

The root cause of most voltage regulation problems is that there is too much impedance in the power system to properly supply the load. Therefore, the voltage drops too low under heavy.

The options:

1- Add voltage regulation, which boosts apparent VI

2- Add shunt capacitor

3- Add series capacitor to cancel inductive voltage drop IX

4- Change the service transformer to reduce impedance Z

5- Reconductor the lines to change to large size to reduce impedance Z

6- Add static var compensators to serve same purpose as capacitors.

There are a variety of voltage regulation devices in use on utility and industrial power systems. We divide them into three major classes:

- Tap changing transformers.
- Isolation devices with separate voltage regulators.
- Impedance compensation devices, such as capacitors.

Voltage Imbalance

Voltage imbalance is defined as maximum deviation from average of 3 phase voltages or currents divided by average of the three phase voltages or currents expressed in percentage.

Imbalance can be defined using symmetrical components. The ratio of either negative or zero components to the positive sequence component equal = percentage imbalance. Sources of voltage imbalance are single-phase loads on three-phase circuit. It can be also the result of blown fuses in one phase of three-phase capacitor bank.

Harmonics

- **Sinusoidal voltages and currents having frequencies are integer multiples of frequency at which the power system designed. Harmonics distortion originates due to the non-linear characteristics of devices and loads in power system.**
-
- **There are several measures commonly used for indicating the harmonic content of the waveform. One of the most common is total harmonic distortion (THD), which can be calculated for either voltage or current:**

$$THD = \frac{\sqrt{\sum_{h=2}^{h_{max}} V_h^2}}{V_1}$$

-
- **THD is a measure of the effective value of the harmonic components of a distorted waveform, that is, the potential heating value of the harmonics relative to the fundamental.**
- **The rms. value of the total waveform is not the sum of the individual components, but the square root of the sum of the squares. THD is related to the rms value of the waveform as follows:**

$$\text{rms}\sqrt{\sum^{hmax} V_h^2} =$$

- **All circuits containing both capacitances and inductances have one or more natrual frequencies. When one of those frequencies lines up with a frequency that being produced on the power system, resonance can develop in which the voltages and current at that frequency continue to persist at high values.**

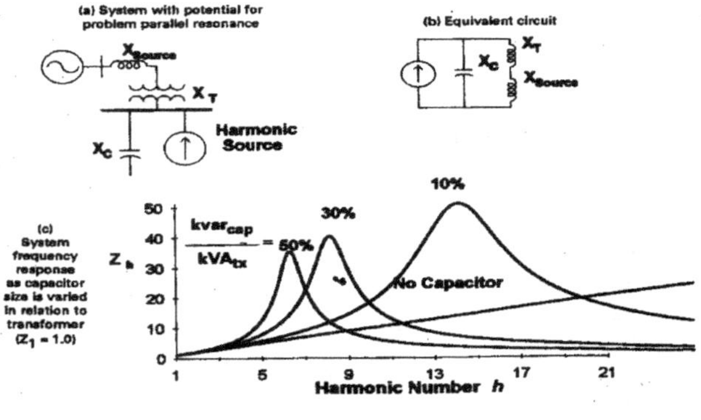

Interharmonics

Voltages or currents having frequency components not integer multiples of frequency at which supply system designed to operate. Sources are i.e. static converters, cycloconverters and induction motors.

Principles for Controlling Harmonics

When the harmonics problem occurs, the basic options for controlling harmonics:

- Reduce the harmonics currents produced by the load: For example adding a line reactor in series with PWM drives will significantly reduce harmonic. Transformer connections can be employed to reduce harmonic in three phase systems. Phase shifting half of the six pulse power converters in a plant by 30 degrees can approximate the benefits of 12-pulse loads by dramatically reducing the fifth and seventh harmonics. Delta connected transformers can block the flow of the zero sequence harmonics from the line. Zigzag and grounding transformers can shunt the triples off the line.
- Add the filters to either siphon the harmonic currents off the system, block the currents from entering the system, or supply the harmonics currents locally: The shunt filter works by short circuiting the harmonics currents close to the source of distortion. This keeps the currents out of supply system. Another approach is to apply a series filter that blocks the harmonic currents. This is a parallel tuned circuit that offers high impedance to harmonic currents. One common application is in the neutral of the grounded wye capacitor to block the flow of triplen harmonics while still retaining a good ground at fundamental frequency.
- Modifying the system frequency response: Adverse system responses to harmonics can be modified by a number of methods: adding a shunt filter, adding a reactor to detune the system, changing the size of the capacitor, moving a capacitor to a point on the system with different short impedance or high losses, removing the capacitor and simply accepting the higher losses.

Filtering

- There are two general classes of filters: Passive filters and Active filters
- Passive filters: Passive filters are made of inductance, capacitance and resistance elements. They are relatively inexpensive compared to other means for eliminating harmonic distortion, but they have the disadvantage of potential adverse with the power system. They are employed either to shunt the harmonic currents off the line or block their flow between parts of the system by tuning the elements to create resonance at the selected frequency. The most common type is the notch series filter. An example of common 480 V arrangement is shown in Fig. 3. One important side effect of adding a filter is that it creates sharp parallel resonance with power system at a frequency below the notch frequency. This resonant frequency must be safely away from any significant harmonic.

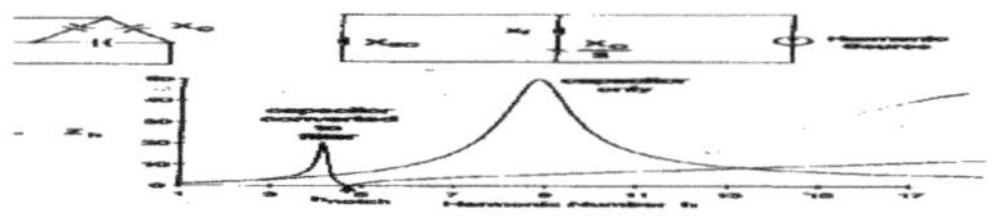

- Active filters: Active filters are based on sophisticated power electronics and are much more expensive than passive filters. However they have distinct advantage that they dont resonate with the system. The basic idea is to replace the portion of the sine wave that is missing in the current in the non-linear load.

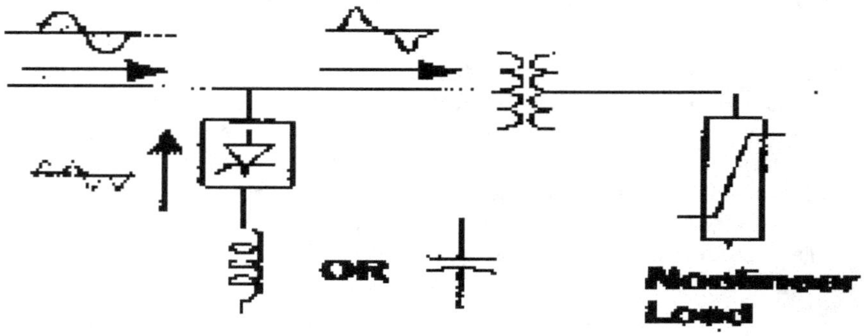

Power Factor of Distorted Waveform

- There are two different types of power factor that must be considered when voltage and current waveforms are not perfectly sinusoidal.
- The first type of power factor is the Input Displacement Factor (IDF) which refers to the cosine of the angle between the 60 Hz voltage and current waveforms.
- The second type is called Distortion Factor (DF) and is defined as follows:
- The Distortion Factor will decrease as the harmonic content goes up.

$$PF = IDF \times DF$$

- Total Power Factor (PF) is the product of the Input Displacement Factor and the Distortion Factor as follows:

$$DF = \frac{1}{\sqrt{1 + THD^2}}$$

Example 1:

Let a Voltage waveform has rms value V and the fundamental and harmonic components are VH_1 =240 V and VH_2=50 V and VH3=40. Then

$$V_{rms} = \sqrt{V_{H1} * V_{H1} + V_{H2} * V_{H2} + V_{H3} * V_{H3}}$$

- $$V_{rms} = \sqrt{240^2 + 50^2 + 40 * 40} = 248.395$$

$$THD = \frac{\sqrt{50^2+40*40}}{240} = 0.267 = 26.7\%$$

- Example 2
- **Use Example 1 data to find PF if IDF=0.9**

$$DF = \frac{1}{\sqrt{1 + THD^2}}$$

$$\text{DF} = \frac{1}{\sqrt{1+0.267^2}}=0.966$$

- PF=DF*IDF=0.9*0.966=0.87

Power Factor in the Presence of Harmonics

- **There are two different types of power factor that must be considered when voltage and current waveforms are not perfectly sinusoidal.**
- **The first type of power factor is the Input Displacement Factor (IDF) which refers to the cosine of the angle between the 60 Hz voltage and current waveforms.**
- **The second type is called Distortion Factor (DF) and is defined as follows:**

$$DF = \frac{1}{\sqrt{1 + THD^2}}$$

- **The Distortion Factor will decrease as the harmonic content goes up.**

Power Factor in the Presence of Harmonics

- **Another method**

$$pf = \frac{I_{power}}{I_{total}}$$

$$I_{total}^2 = I_{power}^2 + I_{reactive}^2 + I_{harmonic}^2$$

Example
For 20 KW motor, find PF
if I_{power}=29 A, $I_{reactive}$=4 A and $I_{harmonic}$=0 A

$$I_{total} = \sqrt{I_{power}^2 + I_{reactive}^2 + I_{harmonic}^2} = \sqrt{29^2 + 4^2 + 0^2} = 29.275$$

Variable Load on Power Stations

- The load on a power station varies from time to time due to uncertain demands of the consumers and is known as variable load on the station.
- The result is that load on the power station varies from time to time.

Effects of variable load.

(1) Need of additional equipment.

- The variable load on a power station necessitates to have additional equipment.
- For instance, if the power demand on the plant increases, it must be followed by the increased flow of coal, air and water to the boiler in order to meet the increased demand. Therefore, additional equipment has to be installed to accomplish this job.

(2) Increase in production cost.

- The variable load on the plant increases the cost of the production of electrical energy.
- The use of a number of generating units increases the initial cost per kW of the plant capacity as well as floor area required. This leads to the increase in production cost of energy.

Daily Load Curve

- The load variations during the whole day (i.e., 24 hours) are recorded half-hourly or hourly and are plotted against time on the graph.
- The curve thus obtained is known as daily load curve as it shows the variations of load w.r.t. time during the day.

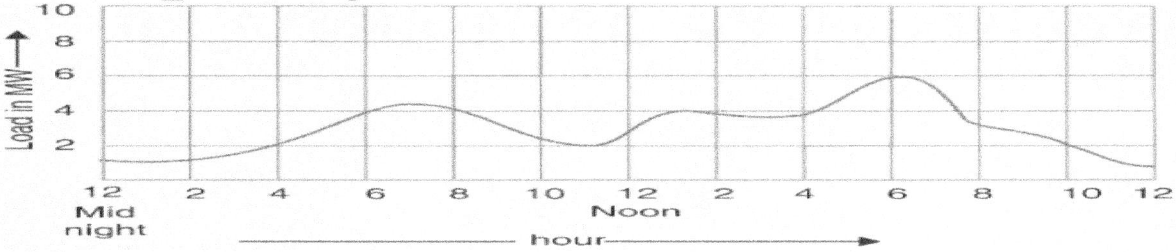

3

Monthly load Curves

- The monthly load curve can be obtained from the daily load curves of that month.
- For this purpose, average values of power over a month at different times of the day are calculated and then plotted on the graph.
- The monthly load curve is generally used to fix the rates of energy.

4

Yearly Load Curves

- The *yearly load curve* is obtained by considering the monthly load curves of that particular year.
- The yearly load curve is generally used to determine the annual load factor.

5

Daily Load Curves

- The area under the daily load curve gives the number of units generated in the day.

  Units generated/day = Area (in kWh) under daily load curve.
- The highest point on the daily load curve represents the maximum demand on the station on that day.
- The area under the daily load curve divided by the total number of hours gives the average load on the station in the day.
- The ratio of the area under the load curve to the total area of rectangle in which it is contained gives the load factor.
- The load curve helps in selecting the size and number of generating units.
- The load curve helps in preparing the operation schedule of the station. The highest point on the daily load curve represents the maximum demand on the station on that day.

6

Important Terms and Factors

1- Connected load:

- It is the sum of continuous ratings of all the equipments connected to supply system.
- For instance, if a consumer has connections of five 100-watt lamps and a power point of 500 watts, then connected load of the consumer is 5 × 100 + 500 = 1000 watts.
- The sum of the connected loads of all the consumers is the connected load to the power station.

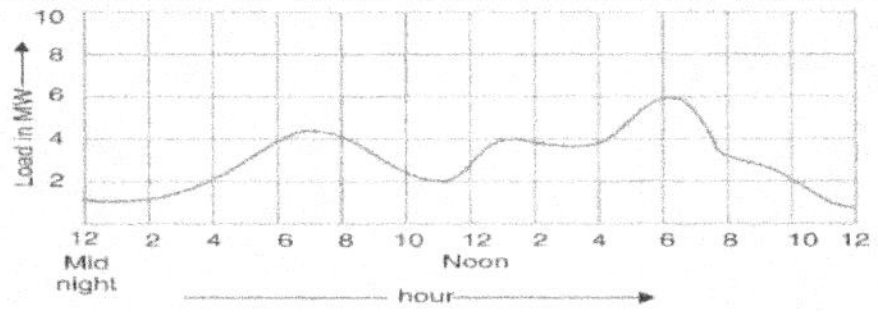

2-Maimum Demand:

- It is the greatest demand of load on the power station during a given period.
- The knowledge of maximum demand is very important as it helps in determining the installed capacity of the station

3-Demand Factor:

- Demand factor is the ratio of maximum demand on the power station to its connected load

$$\text{Demand Factor} = \frac{\text{Maximum demand}}{\text{Connected load}}$$

- The value of demand factor is usually less than 1. If the maximum demand on the power station is 80 MW and the connected load is 100 MW, then demand factor = 80/100 = 0·8.
- The knowledge of demand factor is vital in determining the capacity of the plant equipment.

4- Average load. The area in (kWh) under the load curve on the power station in a given period of time (day or month or year) is known as average load or average demand.

$$\text{Average Load} = \frac{\text{Area (in kWh) under load curve in T time}}{\text{T time}}$$

$$\text{Daily average load} = \frac{\text{No. of units (kWh) generated in a day}}{\text{24 hours}}$$

$$\text{Monthly average load} = \frac{\text{No. of units (kWh) generated in a month}}{\text{Number of hours in a month}}$$

$$\text{Yearly average load} = \frac{\text{No. of units (kWh) generated in a year}}{\text{8760 hours}}$$

5- Load factor. The ratio of average load to the maximum demand during a given period is known as load factor.

$$\text{Load Factor} = \frac{\text{Average load}}{\text{Max. demand}}$$

- The load factor may be daily load factor, monthly load factor or annual load factor if the time period considered is a day or month or year.
- Load factor is always less than 1 because average load is smaller than the maximum demand.
- The load factor plays key role in determining the overall cost per unit generated.
- Higher the load factor of the power station, lesser will be the cost per unit generated.

Example 3.10. *A generating station has the following daily load cycle :*

Time (Hours)	*0—6*	*6—10*	*10—12*	*12—16*	*16—20*	*20—24*
Load (M W)	*40*	*50*	*60*	*50*	*70*	*40*

Draw the load curve and find (i) maximum demand (ii) units generated per day (iii) average load and (iv) load factor.

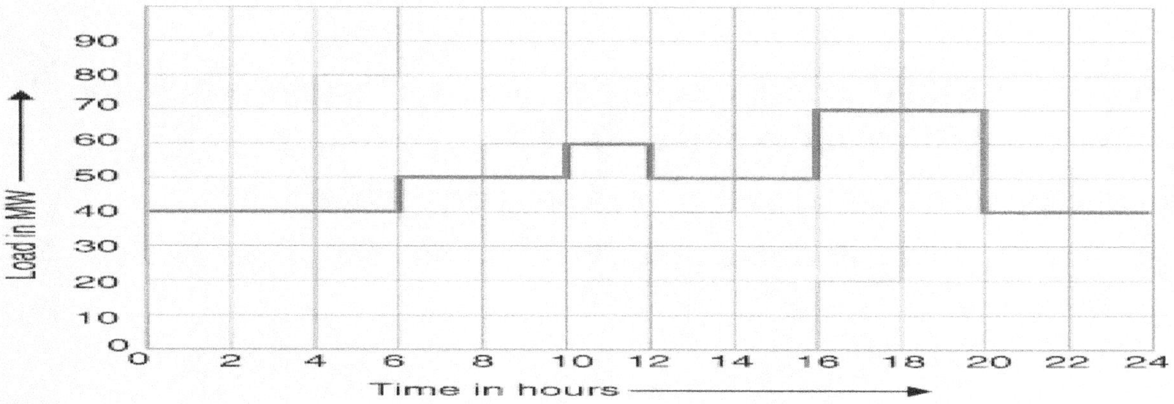

11

Time (Hours)	0--6	6-10	10 -12	12-16	16-20	20-24
Load (MW)	30	60	70	50	90	40

Solution. Daily curve is drawn by taking the load along *Y*-axis and time along *X*-axis. For the given load cycle, the load curve is shown in Fig. 3.6.

(*i*) It is clear from the load curve that maximum demand on the power station is 70 MW and occurs during the period 16—20 hours.

∴ Maximum demand = 70 MW

(*ii*) Units generated/day = Area (in kWh) under the load curve

$$= 10^3\,[40 \times 6 + 50 \times 4 + 60 \times 2 + 50 \times 4 + 70 \times 4 + 40 \times 4]$$

$$= 10^3\,[240 + 200 + 120 + 200 + 280 + 160]\text{ kWh}$$

$$= 12 \times 10^5\text{ kWh}$$

(*iii*) $$\text{Average load} = \frac{\text{Units generated / day}}{24\text{ hours}} = \frac{12 \times 10^5}{24} = 50{,}000\text{ kW}$$

(*iv*) $$\text{Load factor} = \frac{\text{Average load}}{\text{Max. demand}} = \frac{50{,}000}{70 \times 10^3} = 0{\cdot}714 = 71{\cdot}4\%$$

12

6-Diversity factor. The ratio of the sum of individual maximum demands to the maximum demand on power station is known as c

$$\text{Diversity Factor} = \frac{\text{Sum of individual max. demands}}{\text{Max. demand on power station}}$$

- **The maximum demand on the power station is always less than the sum of individual maximum demands of the consumers.**
- **Obviously, diversity factor will always be greater than 1.**
- **The greater the diversity factor, the lesser is the cost of generation of power.**

Example 3

Daily demands of three consumers are given below:

Time	*Consumer 1*	*Consumer 2*	*Consumer 3*
12 midnight to 8 A.M.	*No load*	*200 W*	*No load*
8 A.M. to 2 P.M.	*600 W*	*No load*	*200 W*
2 P.M. to 4 P.M.	*200 W*	*1000 W*	*1200 W*
4 P.M. to 10 P.M.	*800 W*	*No load*	*No load*
10 P.M. to midnight	*No load*	*200 W*	*200 W*

Plot the load curve and find (i) maximum demand of individual consumer (ii) load factor of individual consumer (iii) diversity factor and (iv) load factor of the station.

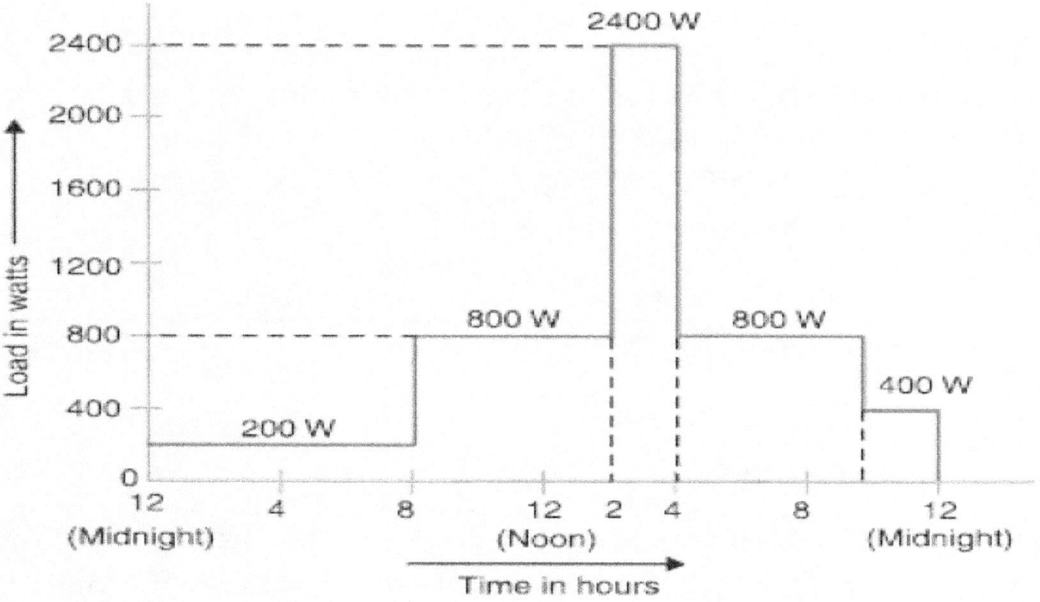

(*i*) Max. demand of consumer 1 = 800 W

Max. demand of consumer 2 = 1000 W

Max. demand of consumer 3 = 1200 W

(*ii*)
$$\text{L.F. of consumer 1} = \frac{\text{Energy consumed / day}}{\text{Max. demand} \times \text{Hours in a day}} \times 100$$
$$= \frac{600 \times 6 + 200 \times 2 + 800 \times 6}{800 \times 24} \times 100 = 45{\cdot}8\%$$
$$\text{L.F. of consumer 2} = \frac{200 \times 8 + 1000 \times 2 + 200 \times 2}{1000 \times 24} \times 100 = 16{\cdot}7\%$$
$$\text{L.F. of consumer 3} = \frac{200 \times 6 + 1200 \times 2 + 200 \times 2}{1200 \times 24} \times 100 = 13{\cdot}8\%$$

(*iii*) The simultaneous maximum demand on the station is 200 + 1000 + 1200 = 2400 W and occurs from 2 P.M. to 4 P.M.

$$\therefore \quad \text{Diversity factor} = \frac{800 + 1000 + 1200}{2400} = 1{\cdot}25$$

(*iv*)
$$\text{Station load factor} = \frac{\text{Total energy consumed / day}}{\text{Simultaneous max. demand} \times 24} \times 100$$
$$= \frac{8800 + 4000 + 4000}{2400 \times 24} \times 100 = 29{\cdot}1\%$$

7-Plant capacity factor:

It is the ratio of actual energy produced to the maximum possible energy that could have been produced during a given period.

$$\text{Plant capacity factor} = \frac{\text{Actual energy produced}}{\text{Max. energy that could been produced}}$$

$$= \frac{\text{Average demand x T}}{\text{Plant capacity x T}}$$

$$\text{Annual plant capacity factor} = \frac{\text{Actual kWh output}}{\text{Plant capacity x 8760}}$$

$$= \frac{\text{Average demand}}{\text{Plant capacity}}$$

- **The plant capacity factor is an indication of the reserve capacity of the plant.**
- **A power station is so designed that it has some reserve capacity for meeting the increased load demand in future.**

Reserve capacity = Plant capacity - Max. demand

8-Plant use factor:

It is ratio of kWh generated to the product of plant capacity and the number of hours for which the plant was in operation.

Example 4

Suppose a plant having installed capacity of 20MW produces annual output of 7.35 x 10^6 kWh and remains in operation for 2190 hours in a year. Then

$$\text{Plant use factor} = \frac{\text{Station output in kWh}}{\text{Plant capacity x Hours of use}}$$

$$\text{Plant use factor} = \frac{7.35 \times 10^6}{(20 \times 10^3) \times 2190} = 0.167 = 16.7\%$$

Example 3.4. *A generating station has a maximum demand of 25MW, a load factor of 60%, a plant capacity factor of 50% and a plant use factor of 72%. Find (i) the reserve capacity of the plant (ii) the daily energy produced and (iii) maximum energy that could be produced daily if the plant while running as per schedule, were fully loaded.*

Solution.

(i) Load factor $= \dfrac{\text{Average demand}}{\text{Maximum demand}}$

or $0{\cdot}60 = \dfrac{\text{Average demand}}{25}$

$\therefore$ Average demand $= 25 \times 0{\cdot}60 = 15$ MW

Plant capacity factor $= \dfrac{\text{Average demand}}{\text{Plant capacity}}$

$\therefore$ Plant capacity $= \dfrac{\text{Average demand}}{\text{Plant capacity factor}} = \dfrac{15}{0{\cdot}5} = 30$ MW

$\therefore$ Reserve capacity of plant = Plant capacity − maximum demand

$= 30 - 25 = 5$ MW

(ii) Daily energy produced = Average demand × 24

$= 15 \times 24 = 360$ MWh

(iii) Maximum energy that could be produced

$= \dfrac{\text{Actual energy produced in a day}}{\text{Plant use factor}}$

$= \dfrac{360}{0 \cdot 72} = 500$ MWh/day

8-Units generated per annum:

It is often required to find the kWh generated per annum from maximum demand and load factor. The procedure is as follows :

$$\textbf{Load factor} = \frac{\textbf{Average load}}{\textbf{Max. demand}}$$

$$\textbf{Average load} = \textbf{Max. demand} \times \textbf{L.F.}$$

Units generated/annum = **Average load (in kW) × Hours in a year**

= **Max. demand (in kW) × L.F. × 8760**

Types of Loads

- The load may be resistive (*e.g.,* electric lamp), inductive (*e.g.,* induction motor), capacitive or some combination of them.
- The various types of loads on the power system are:

1) Domestic load.

- Domestic load consists of lights, fans, refrigerators, heaters, television, small motors for pumping water etc.
- Most of the residential load occurs only for some hours during the day (*i.e.,* 24 hours) *e.g.,* lighting load occurs during night time and domestic appliance load occurs for only a few hours.
- For this reason, the load factor is low (10% to 12%).

2) Commercial load.

- Commercial load consists of lighting for shops, fans and electric appliances used in restaurants etc.
- This class of load occurs for more hours during the day as compared to the domestic load.
- The commercial load has seasonal variations due to the extensive use of air conditioners and space heaters.

3) Industrial load.

- Industrial load consists of load demand by industries.
- The magnitude of industrial load depends upon the type of industry.
- Thus small scale industry requires load up to 25 kW, medium scale industry between 25kW and 100 kW and large-scale industry requires load above 500 kW.
- Industrial loads are generally not weather dependent.

4) Municipal load.

- **Municipal load consists of street lighting, power required for water supply and drainage purposes.**
- **Street lighting load is practically constant throughout the hours of the night.**
- **For water supply, water is pumped to overhead tanks by pumps driven by electric motors.**
- **Pumping is carried out during the off-peak period, usually occurring during the night. This helps to improve the load factor of the power system.**

5) Irrigation load.

This type of load is the electric power needed for pumps driven by motors to supply water to fields. Generally this type of load is supplied for 12 hours during night.

6) Traction load.

This type of load includes tram cars, trolley buses, railways etc. This class of load has wide variation. During the morning hour, it reaches peak value because people have to go to their work place. After morning hours, the load starts decreasing and again rises during evening since the people start coming to their homes.

Typical Demand and Diversity Factor

The demand factor and diversity factor depend on the type of load and its magnitude.

TYPICAL DEMAND FACTORS

Type of consumer		Demand factor
Residence lighting	$\frac{1}{4}$ kW	1·00
	$\frac{1}{2}$ kW	0·60
	Over 1 kW	0·50
Commercial lighting	Restaurants	0·70
	Theatres	0·60
	Hotels	0·50
	Schools	0·55
	Small industry	0·60
	Store	0·70
General power service	0–10 H.P.	0·75
	10–20 H.P.	0·65
	20–100 H.P.	0·55
	Over 100 H.P.	0·50

Typical Demand and Diversity Factor

- **The demand factor and diversity factor depend on the type of load and its magnitude.**

TYPICAL DIVERSITY FACTORS

	Residential lighting	*Commercial lighting*	*General power supply*
Between consumers	3 – 4	1·5	1·5
Between transformers	1·3	1·3	1·3
Between feeders	1·2	1·2	1·2
Between substations	1·1	1·1	1·1

Load Curves and Selection of Generating Units

- **The selection of the number and sizes of the units is decided from the annual load curve of the station.**
- ***The number and size of the units are selected in such a way that they correctly fit the station load curve.***
- **The principle of selection of number and sizes of generating units with the help of load curve is illustrated in the following annual load curve of the station is shown.**
- **It is clear form the curve that load on the station has wide variations ; the minimum load being somewhat near 50 kW and maximum load reaching the value of 500 kW.**
- **The use of a single unit to meet this varying load will be highly uneconomical.**
- **Thus by selecting the proper number and sizes of units, the generating units can be made to operate near maximum efficiency. This results in the overall reduction in the cost of production of electrical energy.**

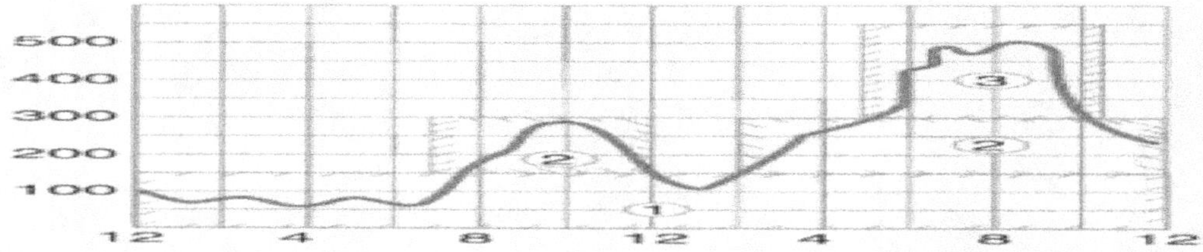

LO2

Explain the role of distribution substation and related equipment

LO2-1

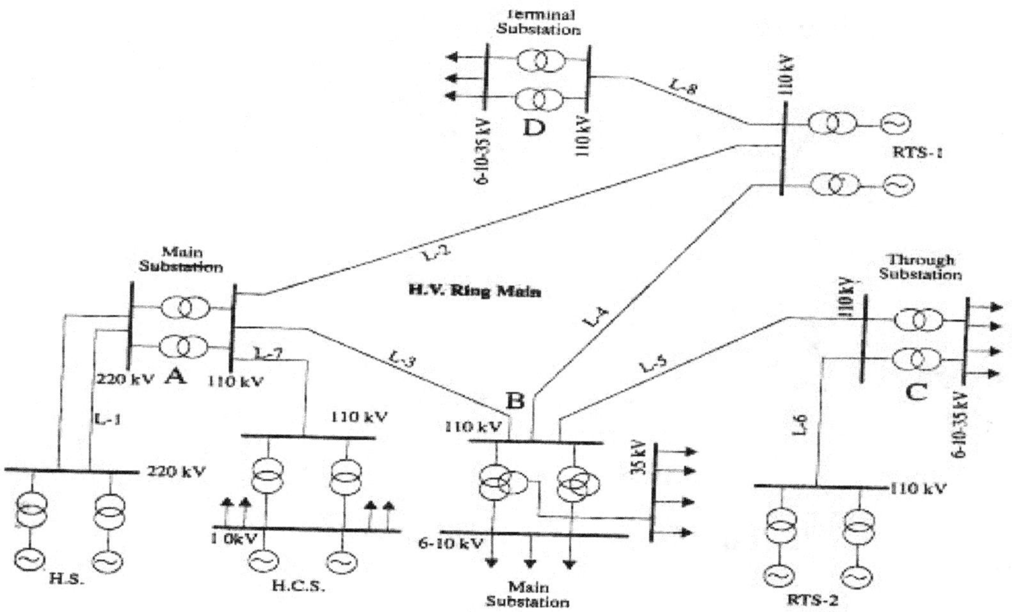

Types of Power System Substations

- A substation is a part of an electrical generation, transmission, and distribution system.
- Substations transform voltage from high to low, or the reverse, or perform any of several other important functions.
- Between the generating station and consumer, electric power may flow through several substations at different voltage levels

Types of Power System Substations

- Generating Station Substation:

This substation steps up the generation voltage (15-23 kV) to the transmission system voltage (69-500kV).

- Transmission Step Down Substation:

This substation interconnects different parts of the transmission system operating at different voltage levels. Examples are: Al Dahma substation & Al Ain South West substation, where power is transformed from 400kV to 220kV.

Types of Power System Substations

- Transmission Switching Substation:

This substation interconnects different parts of transmission system without changing the voltage levels.

- Bulk Power or Grid Substations (secondary transmission):

These substations interconnect transmission (or Sub-transmission) system with the distribution system. They are called Grid Substations and converts power from 220kV to 33kV in Al Ain area. Example of Grid Substation is Zakher Substation.

Types of Power System Substations

- Distribution Substations:

These substations interconnect different parts of the distribution system and it may include transformation of different distribution voltage levels. In Al Ain area, this includes the Primary Substations (33kV/11kV) and the Distribution Substations (11kV to 400V.)

Purpose of Using Substations

Substations do the following functions:

- **Step the voltage level down to a lower voltage levels between two parts of a distribution system (distribution substation).**
- **Isolate a faulty utility component or disconnect a component from the rest of the electric utility system for a scheduled maintenance or repair.**
- **Regulate voltage to compensate for system voltage changes.**
- **Monitor the equipment and circuits operation.**
- **It provides voltage, current and power data for the system operation center.**
- **It also houses the protective devices.**
- **Make interconnections between the electric systems of more than one utility**

Purpose of Using Substations

- **Make interconnections between the electric systems of more than one utility.**
- **Provide electric power sources for reactive power (capacitor banks, synchronous condensers) to improve power factor and voltage control.**

Distribution Substation Equipment

- **The main equipment that can be found in a distribution substation are shown in Figure 1, which shows the single-line diagram of a distribution substation.**

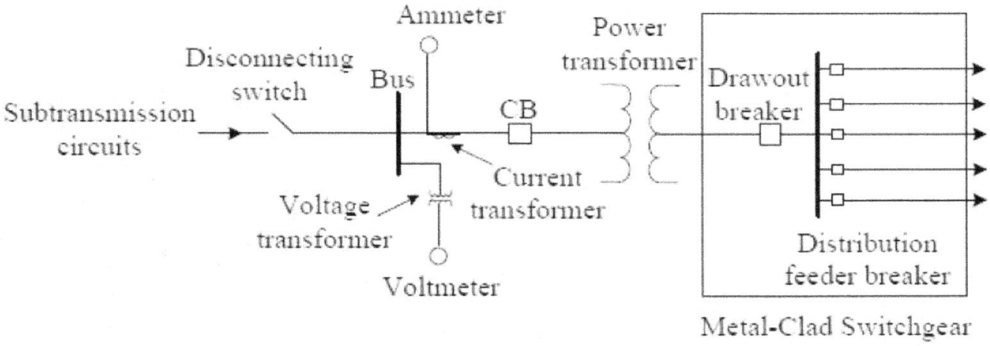

Types of Distribution Substations

1. Normal Substation:

- **It has all the three parts of the substation: High Voltage side (bus work & protection), Power transformers, and Low Voltage side (bus work & protection).**
- **Incoming and outgoing feeders are installed over- or under-ground.**
- **Substation can be in the open air or enclosed in a building. An example of such substation is shown in Figure 2.**

Fig 2: Normal Distribution Substation

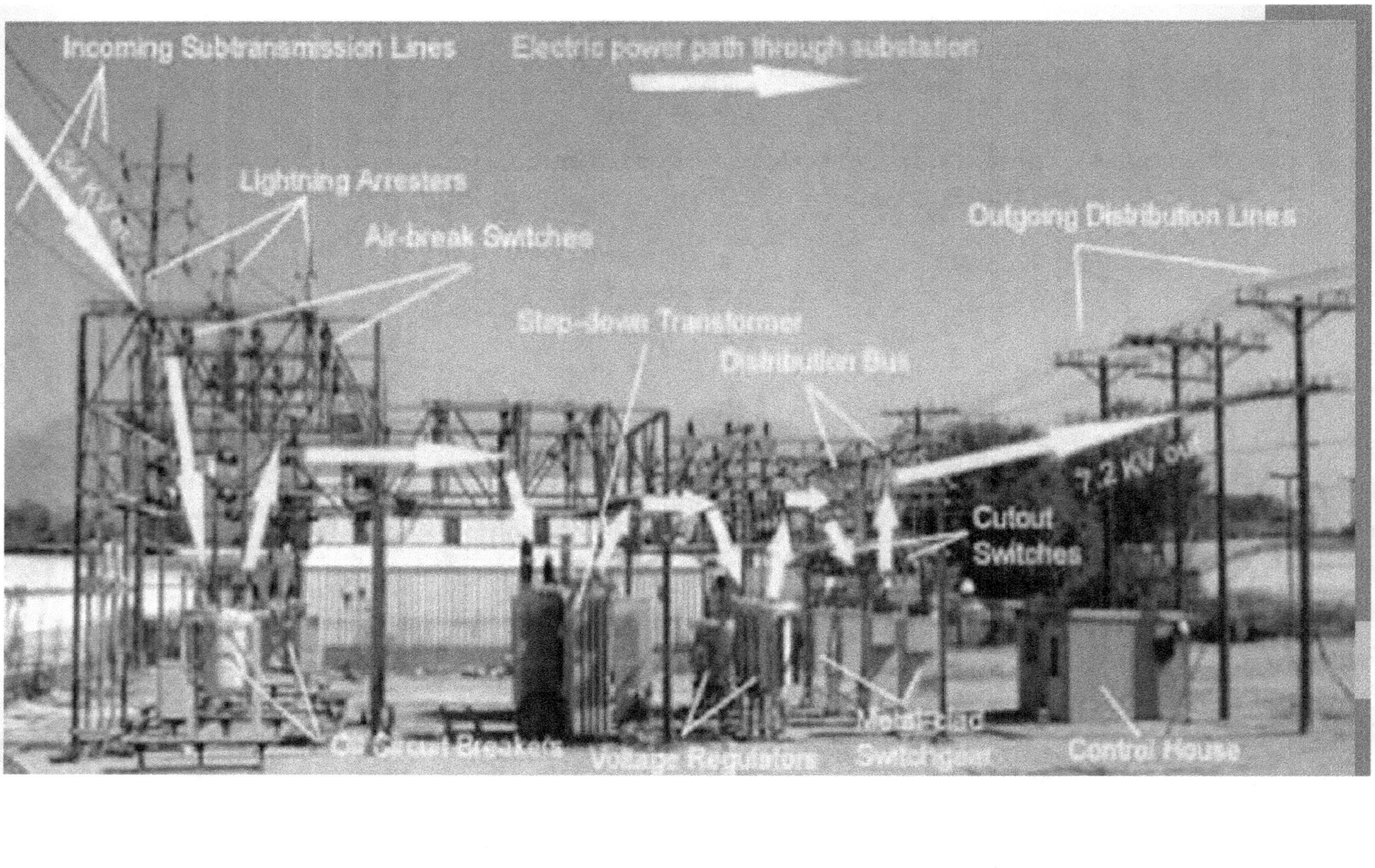

Types of Distribution Substations

2.Substation Unit:

- **This is a substation that has all parts of a normal distribution substation cased in one single package.**
- **It has a metal weatherproof housing that includes the three substation parts separately.**
- **It has its own protection and is constructed as modules.**
- **Figure 3 shows an example of such substation unit.**

Fig 3: Unit Distribution Substation

Types of Distribution Substations

3. Mobile Substations:

- **It is similar to the substation unit but can be moved on large trailer and placed at certain locations near transmission and distribution circuits.**
- **These mobile substations provide maximum reliability and energy continuity following major outages of existing substations.**
- **Its capacity is up to 40 MVA due to size and weight constraints.**
- **It takes from 3-6 hours to be interconnected and energized.**
- **Figure 4 shows an example of such mobile substation.**

Fig 4: Mobile Distribution Substation

Substation Equipment

The following gives a list of equipment used in a typical distribution substation:

- Incoming lines, outgoing lines
- Transformers; main power transformers, auxiliary transformers, CTs and VTs
- HV and LV switchgear, circuit breakers, isolators
- Protection and metering equipment
- SVC equipment including shunt reactors and shunt capacitors
- Surge arrestors, overhead earth wires
- Station earthing grid
- Galvanized steel structures
- substation service equipment such auxiliary station transformer, lighting, air conditioning, auxiliary battery supply and compressed air system
- Control room as well as control and protection panels

Choice of Location of Distribution Substation

The following are some of the factors that need to be considered when deciding the location of distribution substations:

1- Needs to be as close to load center as possible.

2- Proper access to incoming and outgoing lines/cables.

3- Adequate space for expansion.

4- Possibility of land acquisition.

5- Minimum inconvenience to society.

Distribution Substation Types according to Voltage Levels

1. For Voltage levels up to 36 kV:

- Substations for this voltage level are generally of the indoor using metal-clad drawer switchgear.
- Factory assembled substations are nowadays used.
- Switching can be either with Vacuum CBs or SF6 CBs.

17

Distribution Substation Types according to Voltage Levels

2- For Voltage Levels between 36 kV& 66kV:

- Substations for this voltage level are of the outdoor type.
- Switching, metering equipment are usually installed on galvanized steel structures.
- Insulators need regular washing particularly in industrial areas and areas near the sea, trailer mounted substations are also used for remote sites or to provide emergency supplies.

18

Distribution Substation Types according to Voltage Levels

3. For Voltages above 66 kV:

Substations at this voltage level are of THREE types:

a- Conventional Outdoor:

These substations are of the open terminal air-insulated type in which bus bars are visible.

b- Gas Insulated Metal Enclosed (GIS):

Switching, metering equipment in such substations are contained in Aluminum enclosures filled with SF6 gas, for terminal insulation. These substations are generally of the indoor type.

c- Hybrid Substations:

Such substations have equipment of both types of the above mentioned substations (a and b)

Substation Equipment

Every substation, whether indoor or outdoor, high voltage or low voltage has some or all of the equipment described below:

	Equipment	Function
1	Bus-Bars	Various incoming and outgoing circuits are connected to Bus Bars,
2	Circuit Breakers	Switching by automatic interruption of current.
3	Isolators (Disconnecting switches)	Disconnection under no-load conditions.

Substation Equipment

4	Load break switches	Opening circuit under load conditions (load current flowing). Capable of closing under short circuit conditions.
5	Earthing switches	Discharging line voltage to ground safely.
6	Surge arresters	Diverting the high voltage surges to earth and protect insulation.
7	Current Transformers	Stepping down line currents of high voltage lines to safe levels for measurement, protection and control.
8	Voltage Transformers	Stepping down high voltages to safe levels for measurement, protection and control.
9	Series Reactor	Coils used to limit short circuit current levels.
10	Tap-Changing Transformers	Used for voltage control.

Substation Equipment

11	Shunt reactors	Used for long HV transmission lines to control voltage during low load period, by compensating shunt line capacitance.
12	Shunt Capacitors	Power factor improvement, switched during low power factor levels.
13	Static Var Compensator (SVC)	Stepless variation of shunt capacitance.

LO2

Explain the role of distribution substation and related equipment

LO2-2

Overhead Transmission Lines (OHTL) or Underground Cabl

- **Overhead Transmission Lines (OHTL) or Underground Cables used to transmit electricity and provide connection between substations.**

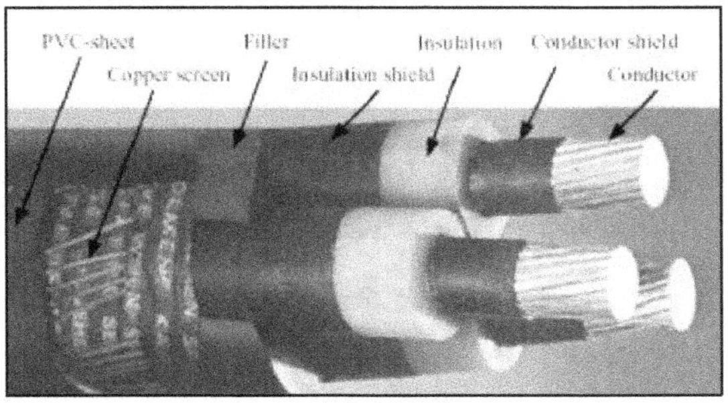

Fig 1. Under Ground Transmission Line

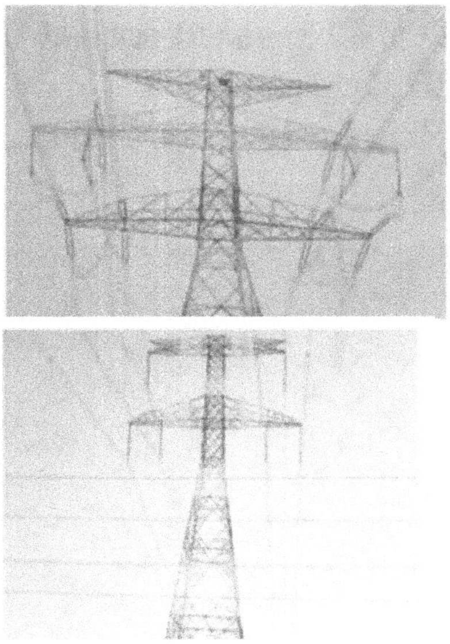

Fig 2. Over Head Transmission Line

- **POWER CABLES**

-Single core-PVC insultaed

-Three core-XLPE insulated-for power flow

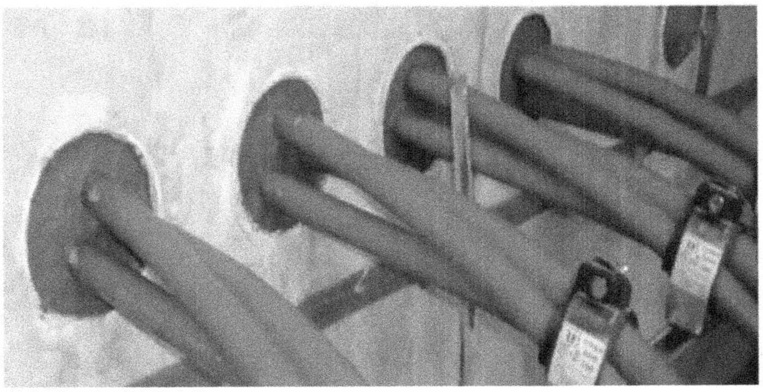

• CONTROL CABLES

-Multi core
-PVC insulated
-Shielded
-For protection, control, measurements etc.

• SURGE ARRESTERS (LIGHTNING ARRESTER)

Surge Arresters discharge the over-voltage surges to earth and protect the equipment insulation from switching surges and lightning surges.

-Connected generally between phase conductor and ground.

-Two types: Gapped Arresters and Gapless Zinc-Oxide Arresters.

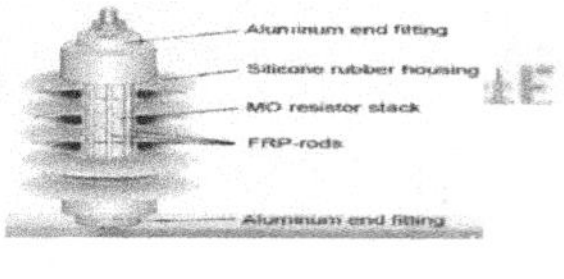

Transformer

Any substation will contain the following:

- Transformer is a device that can convert AC power at one level to AC power of the same frequency at another level.
- The following types of transformer are used in substations:

Substation Transformers: A typical substation would have four Substation Transformers. These transformers are used to step down the voltage from transmission voltages (400kV, 220kV, or 132kV) to distribution voltages (33kV, 22kV, or 11kV).

- **POWER TRANSFORMERS**

To step-up or step-down a.c voltages and to transfer electrical power from one voltage level to another. Tap changers used for voltage control. For very large transformers Transportation, rail permit, etc. should be decided in advance.

-Usually oil filled for outdoor use.

-Three single-phase units to form a three bank used when single three phase unit becomes too large to transport.

-Provided with tap changers

Transformer

- **Auxiliary Transformers: These transformers are used to step down the voltage from the distribution voltages (33kV, 22kV, or 11kV) to 415V in order to energize the substation.**

Fig 3. Sub Station Transformer

Fig 4. Auxiliary Transformer

Transformer

- **Inter Bus Transformers (IBT):** these transformers only exist in a substation that has more than one transmission voltage. It is used to provide electrical connect between (400kV and 220kV) Busbars, (220kV and 132kV) Busbars, or (400kV and 132) Busbars.

- **POWER CABLES**

-Single core-PVC insultaed

-Three core-XLPE insulated-for power flow

- **SHUNT REACTORS**

Used for long EHV transmission line to control voltage during low-load period. To compensate shunt capacitance of transmission line during low load periods.

-Usually oil filled.

-Usually unswitched.

• SHUNT CAPACITORS

For compensating reactive power of lagging power factor. To improve power factor.

-Located at receiving stations and distribution substations.

-Banks rated 132 kV, 66 kV, 33 kV, 11kV, 6.6 kV, etc.

Switched in during heavy loads, switched- off during low loads.

• SERIES CAPACITORS

Used for some long EHV a.c lines to improve power transferability.

-Capacitor bank located at sending –end and or receiving-end of line.

• SERIES REACTORS

To limit short-circuit current and to limit current surges associated with fluctuating loads.
-Located at strategic locations such that fault levels are reduced. Now no more preferred.

Switchgear

- Switchgear is a device used to control the flow of electrical power inside the substation.
- It consists of Busbars, Bus Coupler, Bus Section, isolators, circuit breaker and earth switches.

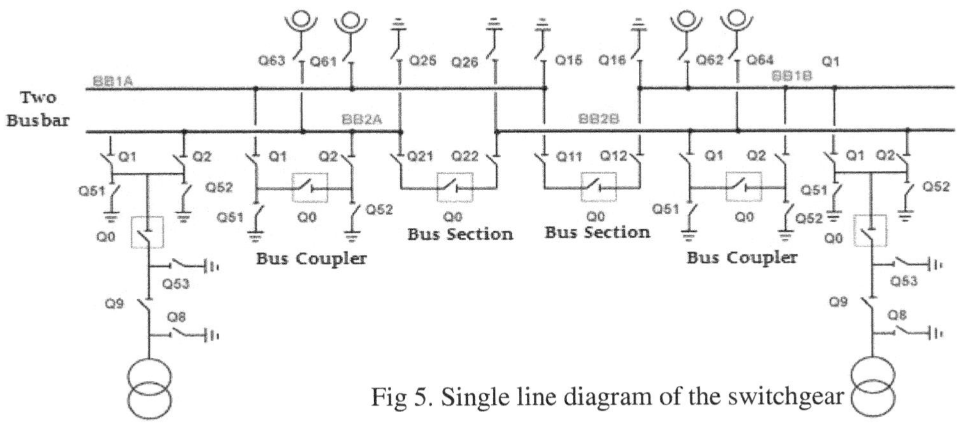

Fig 5. Single line diagram of the switchgear

The following are main components in the switchgear:

1-Busbar:

It is a conductor made of a thick strip of copper or aluminum.

Fig 6. SDBB Substation Bus Bar

Switchgear

2-Busbar Coupler:

- It is used to connect or disconnect two Busbars in the same switchgear.

3-Busbar Section:

- **It is used to connect or disconnect two parts of the same Busbar.**

Fig 7. SDBB Substation Bus Bar Coupler

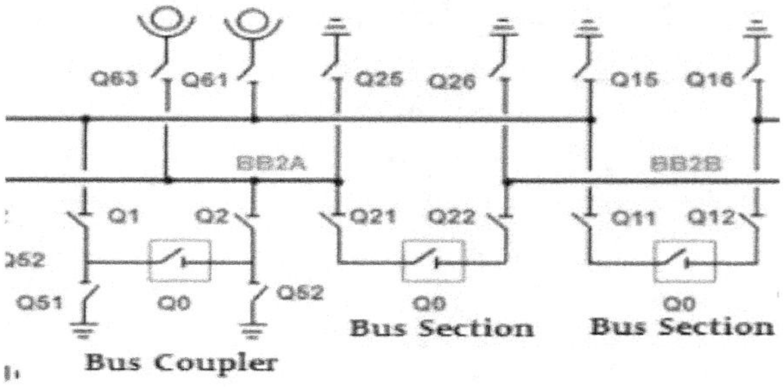

4-Earth Switches:

- Earth Switch is a conductor that is used to discharge electrical charges.
- It protects from accidental operation while working on HV equipment by discharging electrical charge.
- Primary earth is (Q8) and its minimum cross section area is 95mm2.
- Maintenance earth switches are (Q51 & Q52& Q53).
- Busbar earth switches are (Q15 & Q25).

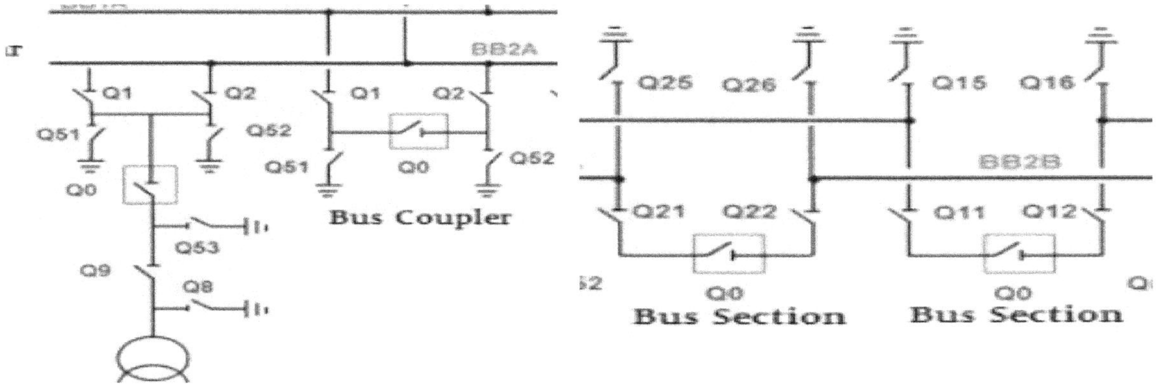

- **EARTHING SWITCH**

Discharging the voltage on the circuit to earth for safety.

-Mounted on the frame of the isolators.

Instrument transformers:

Voltage Transformer (VT):

- **This device is used to step down the voltage in order monitor and measure the voltage in the switchgear.**
- **It provides the required information for the protection system.**

Fig 9. Current Transformer

Current Transformer (CT):

- **This device is used to step down the current in order to measure and monitor the current in the switchgear. It provides the required information for the protection system.**

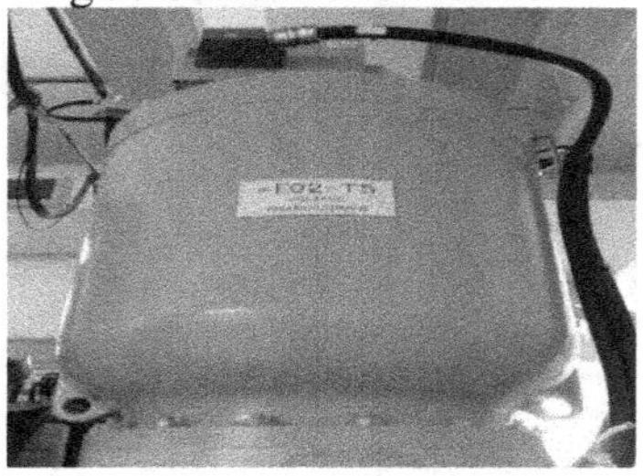

Fig 10. Voltage Transformer

21

- **VOLTAGE TRANSFORMER**

Stepping down voltage for measurement, protection and control.

-Types:

1. Electro magnetic

2. Capacitive (CVT)

- Location on feeder side of circuit breaker.

- **CURRENT TRANSFORMER**

Stepping down current measurement, protection and control.

-Types:

1. Protective CT

2. Measuring CT

-Location decided by protective zone measurement requirements.

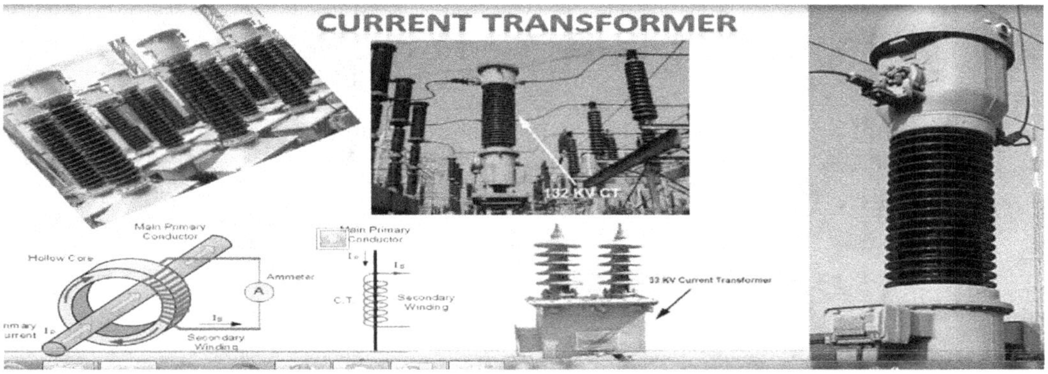

Isolators:

- Isolators are switches that are used to connect or disconnect under no current condition.
- Busbar isolators are (Q1 & Q2)
- Line isolator is (Q9).

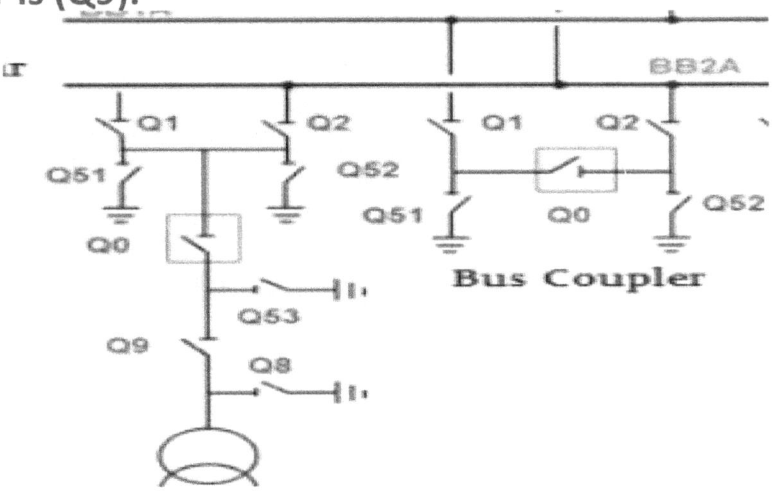

Circuit Breaker (CB):

- **Circuit breaker is a switching device that is used to connect or disconnect circuits in the switchgear.**
- **It is used to serves two basic purposes:**

1. Switch during normal operation conditions for the purpose of operation and maintenance.

2. Switching during abnormal conditions such as short circuit and interrupting the fault currents.

Circuit breaker can operate by different mechanisms system such as:

- **Hydraulic System.**
- **Spring Mechanism.**
- **Hydraulic-Spring system.**
- **Compressed Air.**

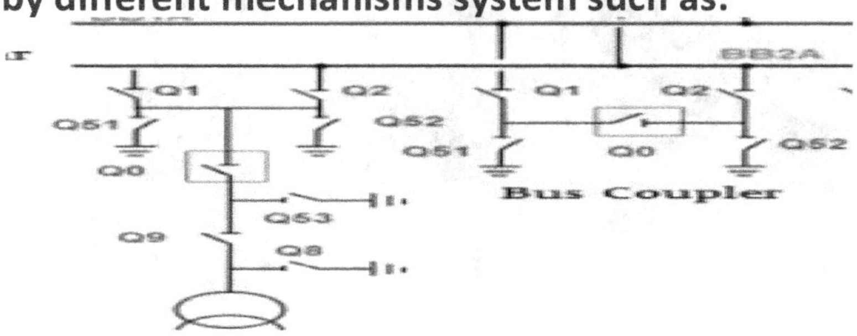

- **The main difference between the circuit breaker and isolator is that CB can interrupt the fault current. It has an interrupting medium for arc quenching.**
- **The interrupting mediums used in CB can be air, oil, vacuum, compressed air or sulfur hexafluoride (SF6).**

Fig 11. A circuit breaker in E18 substation

- **CIRCUIT BREAKER**

Switching during normal and abnormal operating condition. To interrupt short circuit currents.

Operations include:

Closing

Opening

Auto-reclosing

-located near every switching point. Located at both and as every protected zone.

-Types:

Depending on rated voltage: Low voltage, medium voltage, high voltage, extra high voltage.

-Types:

Depending on medium of arc quenching: SF6, Vacuum, Air blast, Minimum Oil

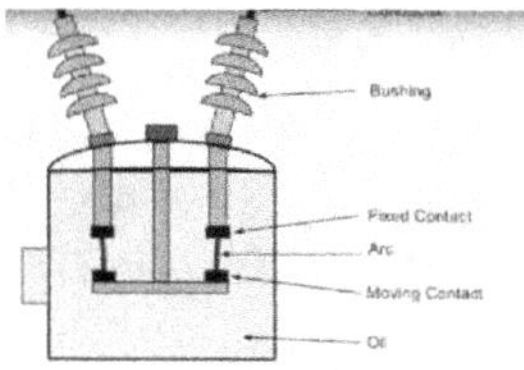

LVAC Room (Low Voltage Alternating Current Room)

- **LVAC room is used to control the flow of low voltage that is needed for the internal use of the substation. Lighting, computers and protection devices are powered from this room.**

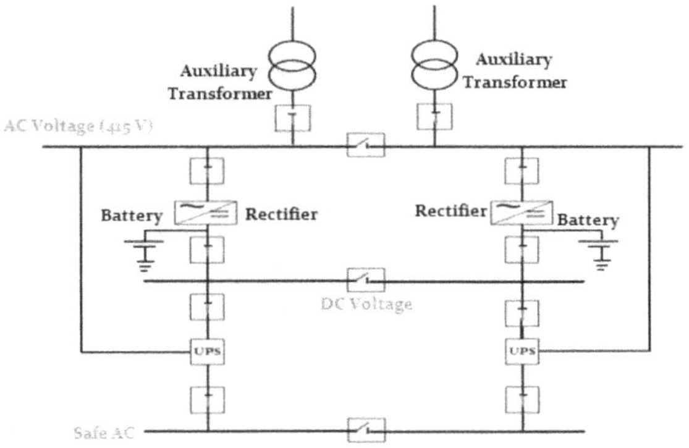

Fig 12. The single line diagram of the LVAC system

LVAC Room (Low Voltage Alternating Current Room)

- In this system, two Auxiliary Transformers are used to step down the voltage from (11kV, 22kV or 33kV) to 415 and feed Busbar 1 & 2.
- Both Busbars are responsible for supplying low voltage AC devices.
- However, some devices, like protection circuits, need DC voltage to operate. Therefore, a rectifier is used to convert the AC voltage to DC.
- In addition, the rectifier works as a charger to charge the 110V& 48V batteries.

Fig 13. LVAC Room

LVAC Room (Low Voltage Alternating Current Room)

- Finally, UPS, Un Interrupting Power Supply, converts DC voltage to safe AC which is used to power that most important systems and devices in the substation.
- If one of the Auxiliary transformers trip, its CB will be open and the Busbar Coupler will be close.
- All the feeders will be connected to the other transformer.
- If both of the transformers tripped, the batteries will energize the substation for approximately 10 hours.

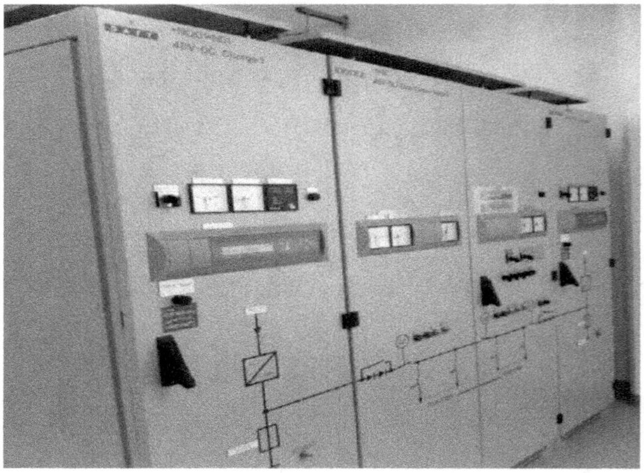

Fig 14. 48V Charger

LVAC Room (Low Voltage Alternating Current Room)

Fig 15. UPS

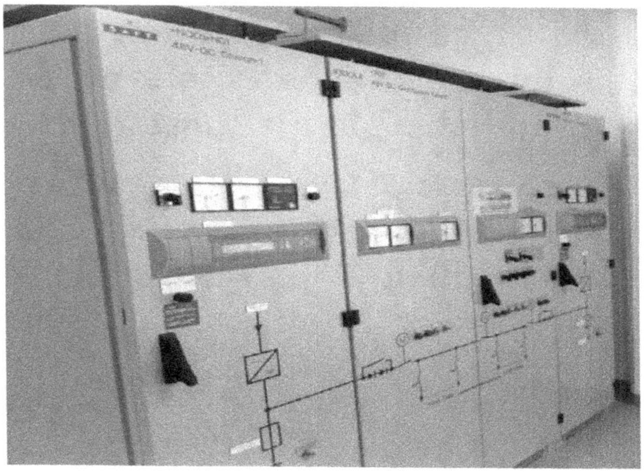

Fig 16. 110V Charger

DC Room (Battery Room)

- In this room, the DC batteries are kept.
- It contains two DC voltage levels which are 110V and 48V.
- 110V batteries are used to supply the protection system and control.
- 48V batteries supplies Telecommunication.
- DC Room also works as a backup source of energy in case any failure in the auxiliary transformers.

Fig 17. DC Room

HVAC Room (Heating, Ventilation and Air Conditioning)

- **As suggested for its name, this room is used to control air conditioning in the substations.**
- **It consists of two systems which are Condensate Unite (CU) and Air Handling Unite (AHU)**

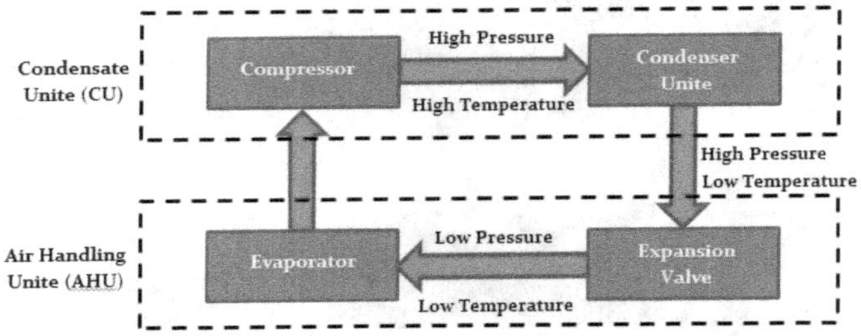

Fig 18. Cooling Process

Relay Room

- All the relays and protection systems of the substation are kept in this room.
- Relays are automatic devices which can sense the fault and then send signal to the circuit breaker to operate (Open).

Fig 19. Relay Room

SCMS Control Room (Substation Control and Monitoring System)

- **This room is used to control and monitor all the equipment in the substation.**

Fig 20. SCMC Room

Telecommunication Room

- **All the communication devices that are needed to connect the substation with Network Management Division are kept in this room.**
- **These devices transmit signals and information about the substation to NMD as well as receive control signals from it.**

Fig 21. Telecommunication Room

The Co2 Room

- The Co2 cylinders that are used to put out fire in the substation are kept in this room

Fig 22. CO2 Room

Firefighting Pump Room

- **This room contains water pumps that are used for firefighting.**
- **The water is provided by water drum and then this room pumps it the needed place to put out the fire.**

Fig 23. Water Drum

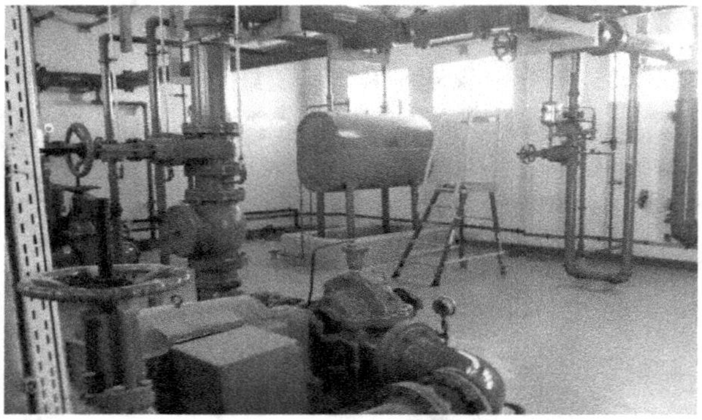

Fig 24. Fire Fighting Pump Room

• MARSHALLING KIOSKS

To mount monitoring instruments, control equipments and to provide access to various transducers.

Control and protective cables are laid between Marshalling Kiosks located I switchyard and corresponding indoor control panels.

-Located in switchyard near every power transforme

• METERING PANELS, CONTROL AND RELAY PANELS

To house various measuring instruments, control instruments, protective relays.

-Located in air-conditioned building. Control cables are laid between switchyard equipment and these panels.

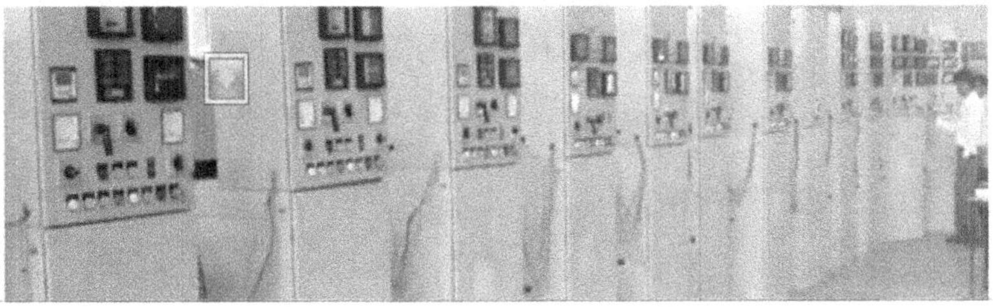

- **MEDIUM AND LOW VOLTAGE A.C SWITCHGEAR**

To provide a.c power to auxiliaries, station-lighting system, etc. at respective voltage levels.
-Located inside switchgear building.

- **STATION EARTHING SYSTEM**

To provide a low resistance earthing for
-discharging currents from surge arresters, earthing switches.
-for equipment body earthing.
-for providing path for neutral to ground currents for earth fault protection
-Earth mat and earth electrodes placed below ground level. Connected to equipment structures, neutral points for the purpose of equipment earthing and neutral point earthing.

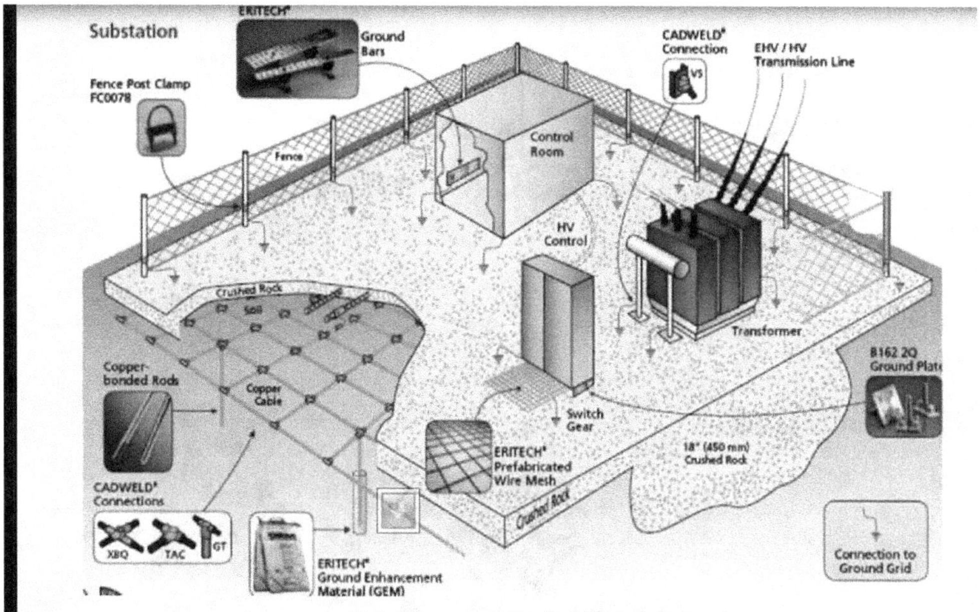

General Arrangement of an Earth Electrode System at an Electrical Sub-Station

- **ISOLATED PHASE BUS SYSTEM**

Provides connection between generator and main transformer.
-Aluminium enclosures for each phase conductor.

• INSULATORS

Porcelain, Glass, epoxy for indoor use.

-String Insulators-For flexible ASCR conductors.

1. Tension

2. Suspension

-Post insulators

-For tubular Conductors.

-For apparatus.

• NEUTRAL GROUNDING EQUIPMENT

To limit short-circuit current during ground fault.

-Short time rated. Connected between neutral point and ground.

- **LINE TRAP**

Inductive coil usually outdoor.

-Connected in outdoor yard incoming line.

LO2

Explain the role of distribution substation and related equipment

LO2-3

Gas Insulated (GIS) Switch Gear

- GIS is a type of switchgears that contains Sulfur Hexafluoride gas (SF6).
- This gas plays an essential role in providing insulation as well as having excellent arc quenching properties.

Gas Insulated (GIS) Switch Gear

- There are many *advantages* behind using GIS.
- The first advantage is compactness in which the switchgear is small and occupies a limited area.
- It has a high degree of safe operation.
- It provides protection against pollution in which all equipments are encapsulated in enclosures and fully protected from environment effects.

Bus-bar Configurations in Distribution Substations

- **Bus Bars are copper bars that operate at constant voltage.**
- **All generators, transformers, feeders etc that operate at the same voltage level are connected to some sort of bas-bar systems in substations.**
- **All bus bars mentioned are three-phase.**

Bus-bar Configurations in Distribution Substations

- Depending of the interconnections of generators, transformers and load feeders, a number of bus bar arrangements are used.
- Arrangement of the switching devices will impact
  - ❖ maintenance
  - ❖ protection,
  - ❖ initial substation development
  - ❖ cost.

Bus-bar Configurations in Distribution Substations

- Some of the more common types of bus bar arrangements are:

1. Single Bus-Bar System

- This is the simplest bus-bar arrangement.
- It is used for direct connection of generators, transformers, feeders. . etc.
- All these equipment are connected to the bus bars through circuit breakers and isolators.

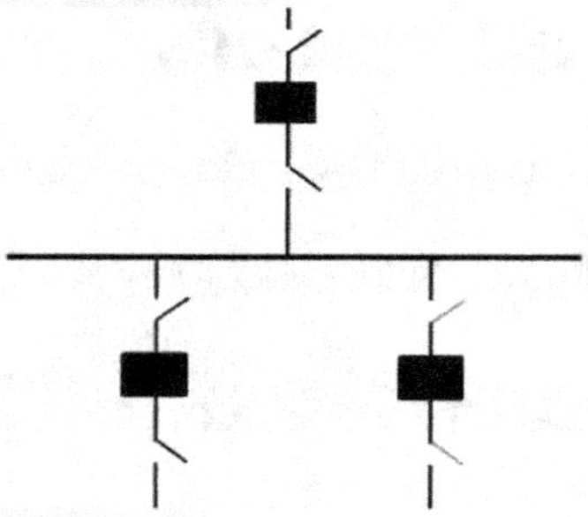

Fig 1. Single Bus Bar System

- This arrangement involves one main bus with all circuits connected directly to the bus.
- The reliability of this type of an arrangement is very low. When properly protected by relaying, a single failure to the main bus or any circuit section between its circuit breaker and the main bus will cause an outage of the entire system.
- In addition, maintenance of devices on this system requires the de-energizing of the line connected to the device.
- Maintenance of the bus would require the outage of the total system, use of standby generation, or switching to adjacent station, if available.
- Since the single bus arrangement is low in reliability, it is not recommended for heavily loaded substations or substations having a high availability requirement.

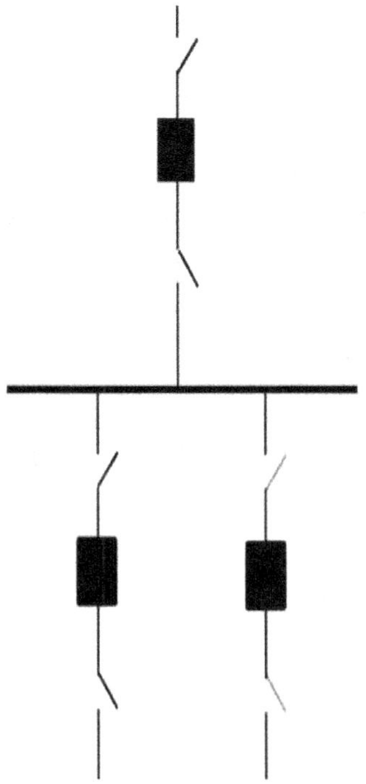

Advantages:

a. Simple design and construction
b. Lowest cost
c. Small land area requirement
d. Easily expandable
e. Needs little maintenance

Disadvantages:

a. Bus bars cannot be cleaned or repaired without de-energizing the whole system.
b. In case of a fault on the bus bar, the entire bus bar has to be de-energized which results in the total system shut down.
c. Lowest reliability

2. Main and Transfer Bus Configuration

- This scheme is arranged with all circuits connected between a main (operating) bus and a transfer bus (also referred to as an inspection bus).
- Some arrangements include a bus tie breaker that is connected between both buses with no circuits connected to it.

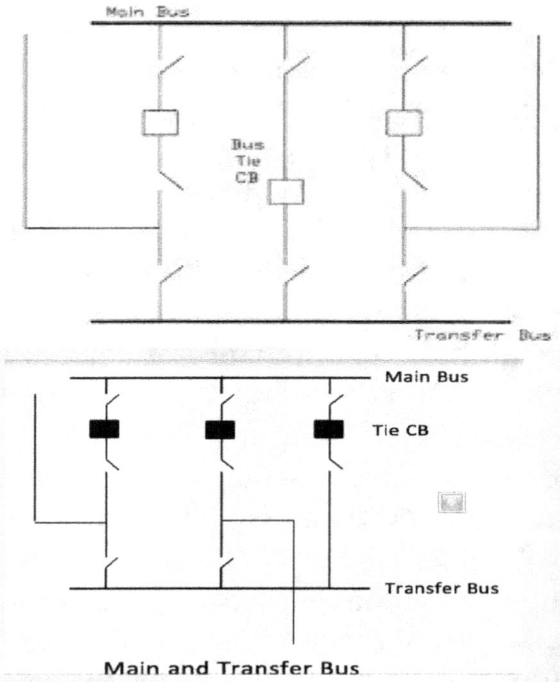

Fig 7. Main and Transfer Bus Bar System

- Since all circuits are connected to the single, main bus, reliability of this system is not very high.
- However, with the transfer bus available during maintenance, de-energizing of the circuit can be avoided.
- Some systems are operated with the transfer bus normally de-energized.
- When maintenance work is necessary, the transfer bus is energized by closing the tie breaker. Then the breaker is taken out of service.
- The circuit breaker remaining in service will now be connected to both circuits through the transfer bus.
- This arrangement is slightly more expensive than the single bus arrangement, but does provide more flexibility during maintenance.
- Protection of this scheme is similar to that of the single bus arrangement.
- The area required for a low profile substation with a main and transfer bus scheme is also greater than that of the single bus, due to the additional switches

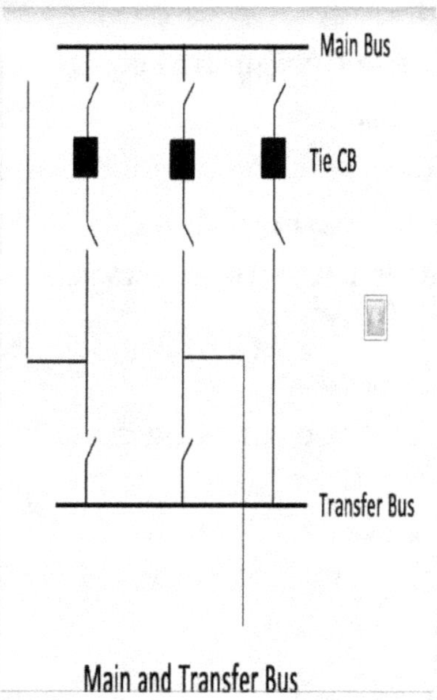

3. Double Bus Bar, Double Breaker

- This scheme provides a very high level of reliability by having two separate breakers available to each circuit.
- In addition, with two separate buses, failure of a single bus will not impact either line.
- Maintenance of a bus or a circuit breaker in this arrangement can be accomplished without interrupting either of the circuits.

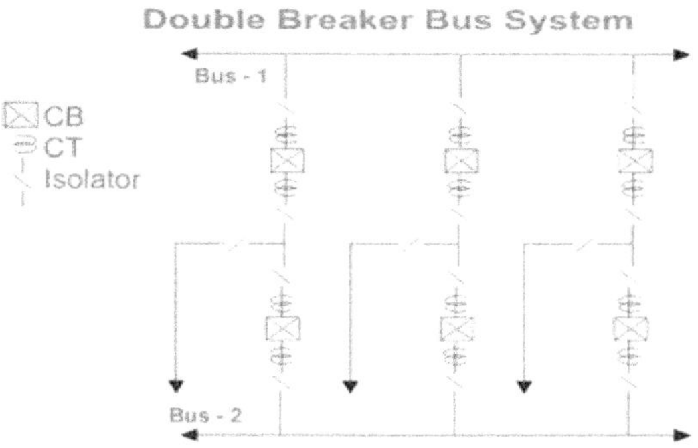

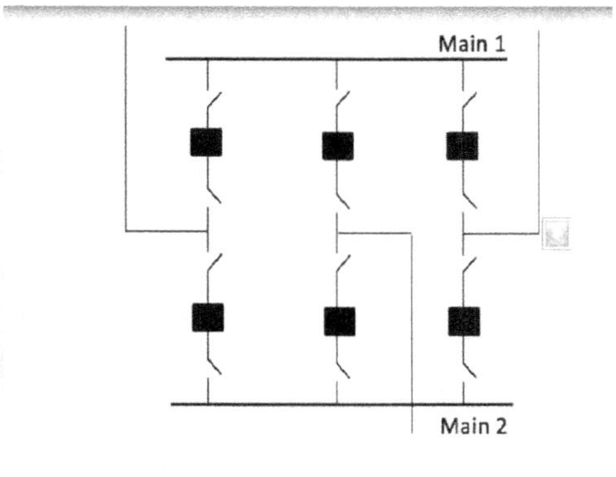

Double Bus Double Breaker

Fig 4. Double Breaker Double Bus Bar System

4. Double Bus-Bar, Single Breaker

- There are two bus bars; the "main bus bar" and the "auxiliary, or reserve, bus bar".
- The bus-coupler shown can be closed so as to connect the two bus bars systems, if needed.
- All units can be connected to either bus bars through circuit breakers and isolators.
- However, operating the bus tie breaker in the normally open position defeats the advantages of the two main buses.
- It arranges the system into two single bus systems, which as described previously, has very low reliability.
- Relay protection for this scheme can be complex, depending on the system requirements, flexibility, and needs.
- With two buses and a bus tie available, there is some ease in doing maintenance, but maintenance on line breakers and switches would still require outside the substation

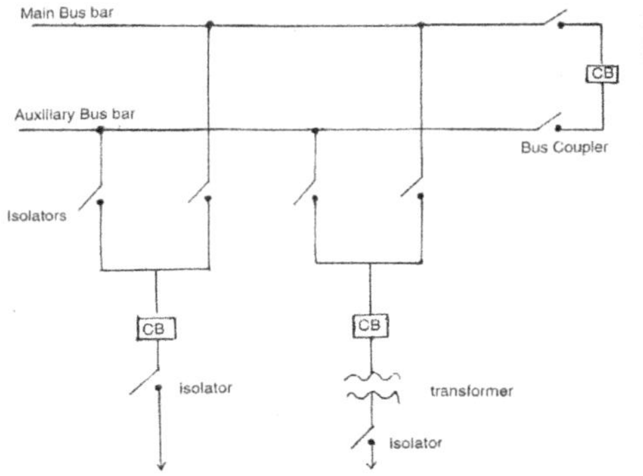

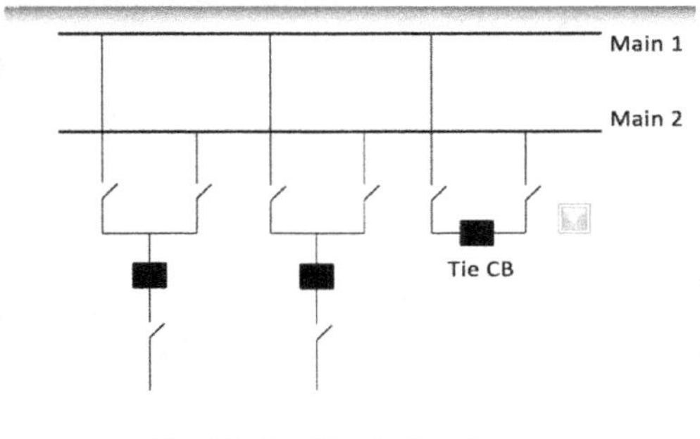

Double Bus Single Breaker

Fig 5. Duplicate Single Bus Bar System

Advantages:

a. If a fault occurs on any bus-bar, it can be isolated and repaired after transferring all units to the other bus-bar.
b. New feeders/units can be connected without disturbance to other feeders/units.
c. Allow periodic maintenance of supply without total shut down.
d. System flexibility and reliability is high.
e. No interruption of service to any circuits from bus fault

Disadvantages:

a. Additional circuit breakers, isolators and transformers are required with each extra section to be connected, making the arrangement more expensive and relatively less reliable.

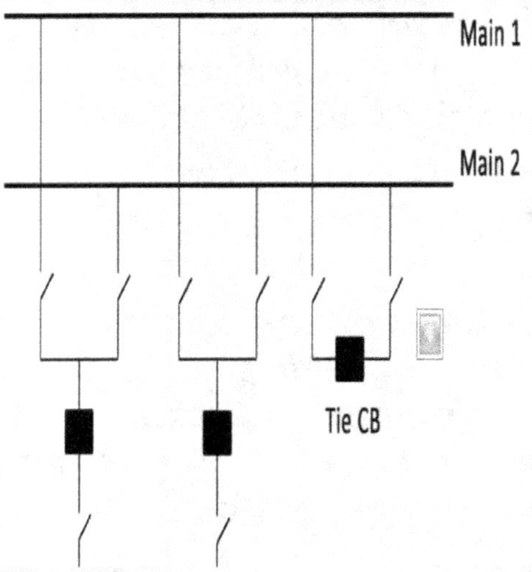

5. Ring Bus-Bar System:

- This system requires only one circuit breaker per line.
- It provides greater flexibility and higher reliability of supply as each feeder is supplied from two ends.
- Arranging the scheme in this manner will minimize the potential for the loss of the supply to the ring bus due to a breaker failure.
- Relaying is more complex in this scheme than some previously identified.
- Since there is only one bus in this scheme, the area required to develop this scheme is less than some of the previously discussed schemes.
- However, expansion of a ring bus is limited, due to the practical arrangement of circuits.

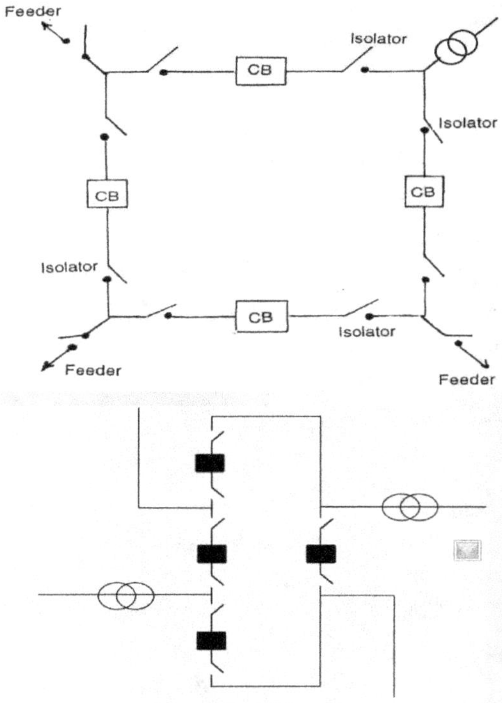

Fig 5. Ring Bus Bar System

- In this scheme, as indicated by the name, all breakers are arranged in a ring with circuits tapped between breakers.
- For a failure on a circuit, the two adjacent breakers will trip without affecting the rest of the system.
- Similarly, a single bus failure will only affect the adjacent breakers and allow the rest of the system to remain energized.
- However, a breaker failure or breakers that fail to trip will require adjacent breakers to be tripped to isolate the fault.

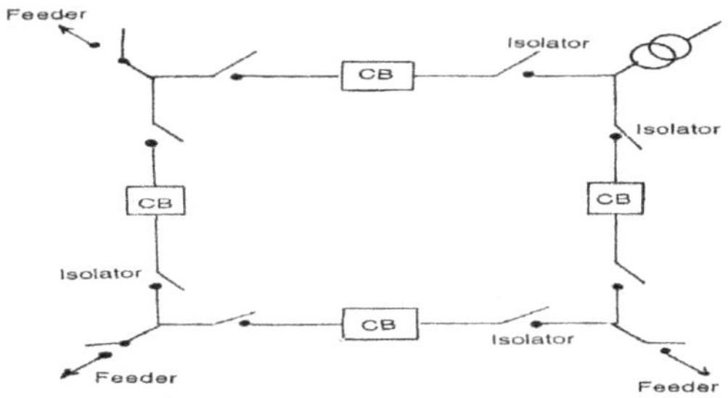

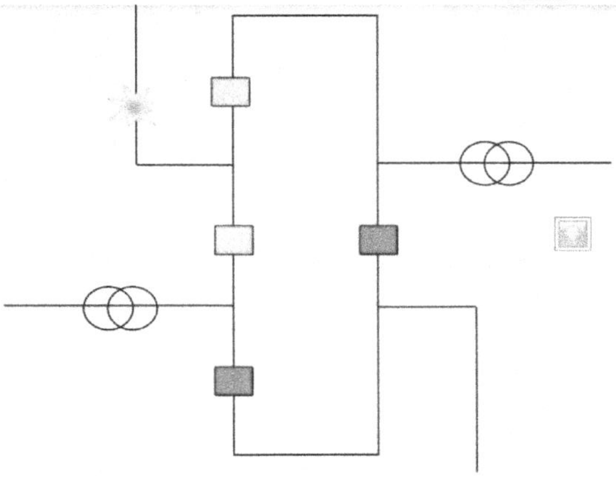

Fig 5. Ring Bus Bar System

- Maintenance on a circuit breaker in this scheme can be accomplished without interrupting any circuit, including the two circuits adjacent to the breaker being maintained.
- The breaker to be maintained is taken out of service by tripping the breaker, then opening its isolation switches.
- Since the other breakers adjacent to the breaker being maintained are in service, they will continue to supply the circuits.
- In order to gain the highest reliability with a ring bus scheme, load and source circuits should be alternated when connecting to the scheme.

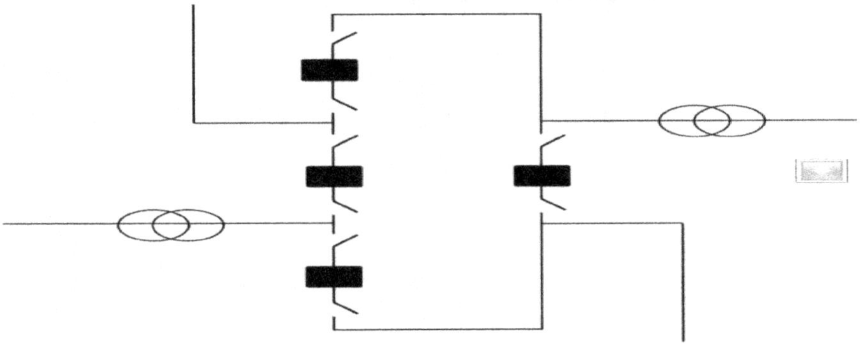

Ring Bus

6. Breaker-and-a- Half Bus Bar System

- **The name of this system is derived from the fact that three circuit breakers are required for every two feeders, or 1.5 breakers per feeder.**
- **This system is usually used with substations with large number of feeders.**

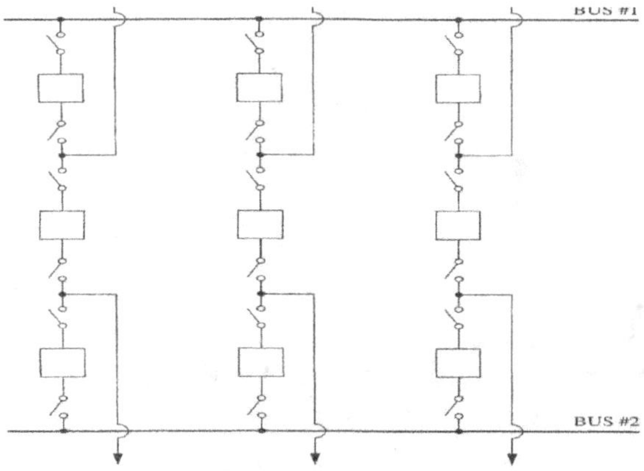

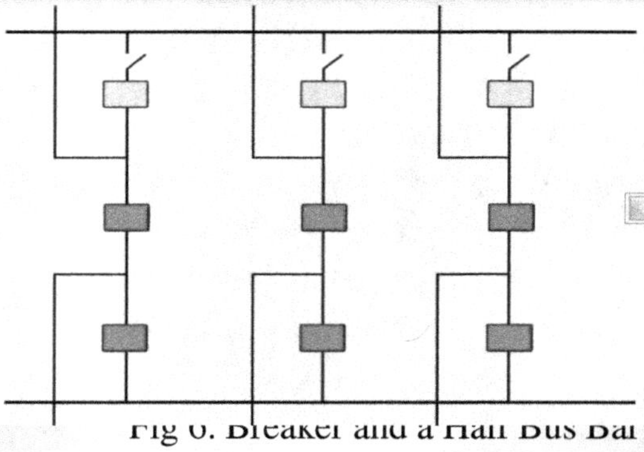

Fig 6. Breaker and a Half Bus Bar System

- **The breaker-and-a-half scheme can be developed from a ring bus arrangement as the number of circuits increases.**
- **In this scheme, each circuit is between two circuit breakers, and there are two main buses.**
- **The failure of a circuit will trip the two adjacent breakers and not interrupt any other circuit.**
- **With the three breaker arrangement for each bay, a center breaker failure will cause the loss of the two adjacent circuits.**
- **However, a breaker failure of the breaker adjacent to the bus will only interrupt one circuit.**
- **Maintenance of a breaker on this scheme can be performed without an outage to any circuit.**
- **Further- more, either bus can be taken out of service with no interruption to the service.**
- **This is one of the most reliable arrangements, and it can continue to be expanded as required.**
- **Relaying is more involved than some schemes previously discussed.**
- **This scheme will require more area and is costly due to the additional components.**

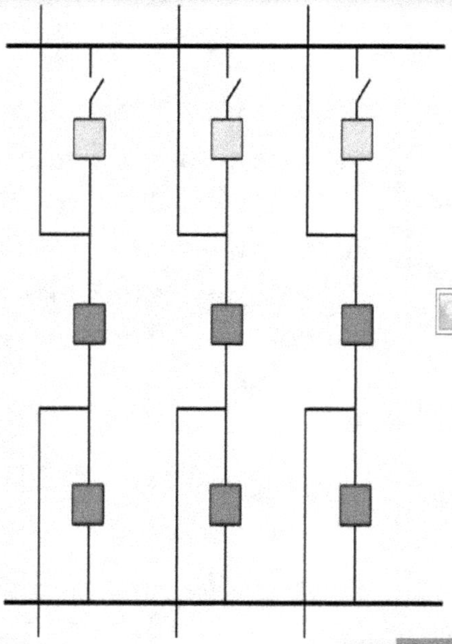

Advantages:

a. **Flexible operation**

b. **High reliability, as a bus or feeder fault can be isolated while the station remains in service.**

c. **Reasonably economical and easy to expand.**

d. **Can isolate either main bus for maintenance without disrupting service**

e. **Double feed to each feeder**

f. **Bus fault does not interrupt service to any circuits**

Disadvantages:

a. **One-and-a-half breakers are required per circuit**

7. Sectionalized Single Bus-Bar System:

- **In large substations, where a large number of units is to be inter-connected using the single bus bar system, the bus bar is divided (sectionalized) into a number of sections through circuit breakers and isolators.**

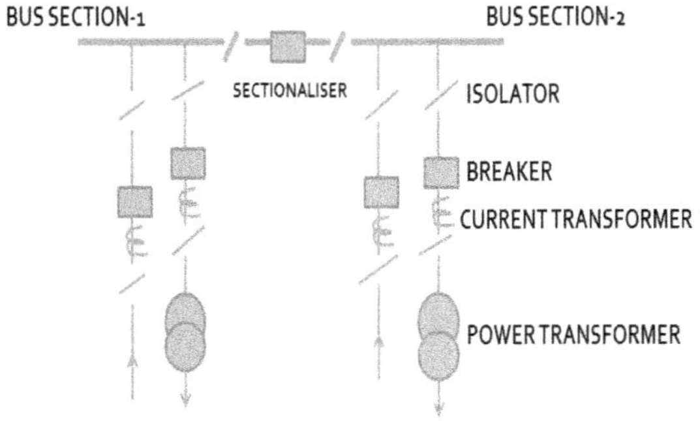

Fig 2. Sectionalized Single Bus Bar System

Advantages:

1. **In case of repair, only the faulty section is to be isolated.**
2. **Loss of only part of the substation for a bus bar fault**
3. **Future expansion is easy.**
4. **Flexible operation**
5. **Increased level of reliability**

Disadvantages:

1. **Additional circuit breakers, isolators and transformers are required with each extra section to be connected, making the arrangement more expensive and relatively less reliable.**

Configuration	Reliability	Cost	Available area
Single bus	Least reliable — single failure can cause complete outage	Least cost — fewer components	Least area — fewer components
Double bus	Highly reliable — duplicated components; single failure normally isolates single component	High cost — duplicated components	Greater area — twice as many components
Main bus and transfer	Least reliable — same as Single bus, but flexibility in operating and maintenance with transfer bus	Moderate cost — fewer components	Low area requirement — fewer components
Double bus, single breaker	Moderately reliable — depends on arrangement of components and bus	Moderate cost — more components	Moderate area — more components
Ring bus	High reliability — single failure isolates single component	Moderate cost — more components	Moderate area — increases with number of circuits
Breaker and a half	Highly reliable — single circuit failure isolates single circuit, bus failures do not affect circuits	Moderate cost — breaker-and-a-half for each circuit	Greater area — more components per circuit

LO2

Explain the role of distribution substation and related equipment

LO2-4

Operational and Safety Switching

Example 1: Overhead Transmission Line (OHTL) & Underground Cable (Feeder):

To isolate an overhead Transmission line and do the operational and safety switching, the following procedure is used:

- **Identify which substation is the source and which is the destination (for example S/S A is the source and S/S B is the destination).**
- **Open circuit breaker (Q0) form the destination side (S/S B).**
- **Open circuit breaker (Q0) form the source side (S/S A).**
- **Open Busbar isolator (Q1) and line isolator (Q9) in S/S A.**
- **Open Busbar isolator (Q1) and line isolator (Q9) in S/S B.**
- **Close the primary earth (Q8) in both of the substations.**
- **Finally, identify Point of Isolation and issue safety document (PTW...Permit to Work) for the competent person.**

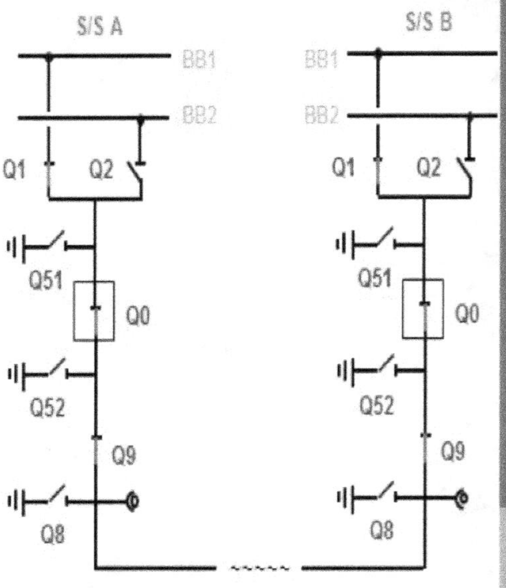

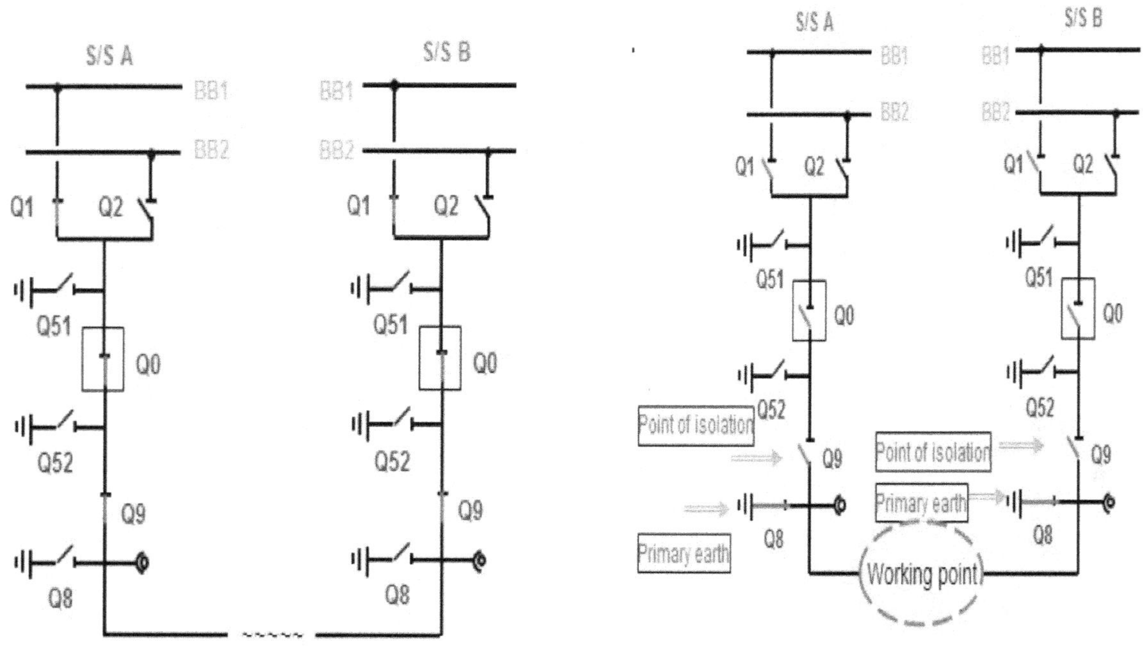

Fig 1. OHL or underground cable before and after doing operational and safety switching

• Example 2: Busbar isolation

To isolate a Bus bar (for example BB1A) and do the operational and safety switching, the following procedure is used:

- **Open Circuit breaker (Q0) of all the Bus Couplers 1(BC1) and Bus Section 1(BS1) in the switchgear.**
- **Transfer all the feeders from BB1A to BB2A by first closing (Q2) of each feeder and then open (Q1).**
- **Check that nothing is connected to BB1A. Then, close BB1A earth (Q15).**
- **Finally, identify Point of Isolation and issue safety document (PTW) for the competent person.**

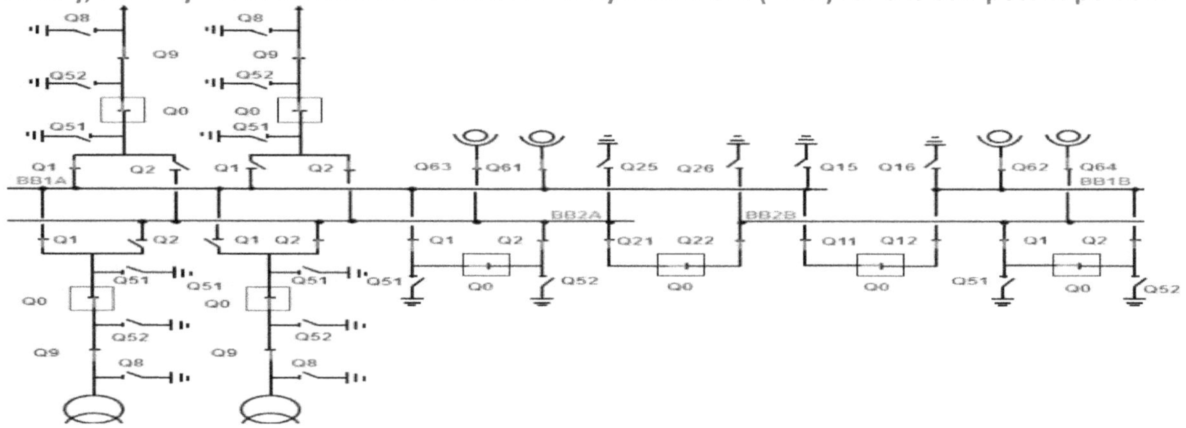

Fig 2. Bus bar before doing operational before and safety switching

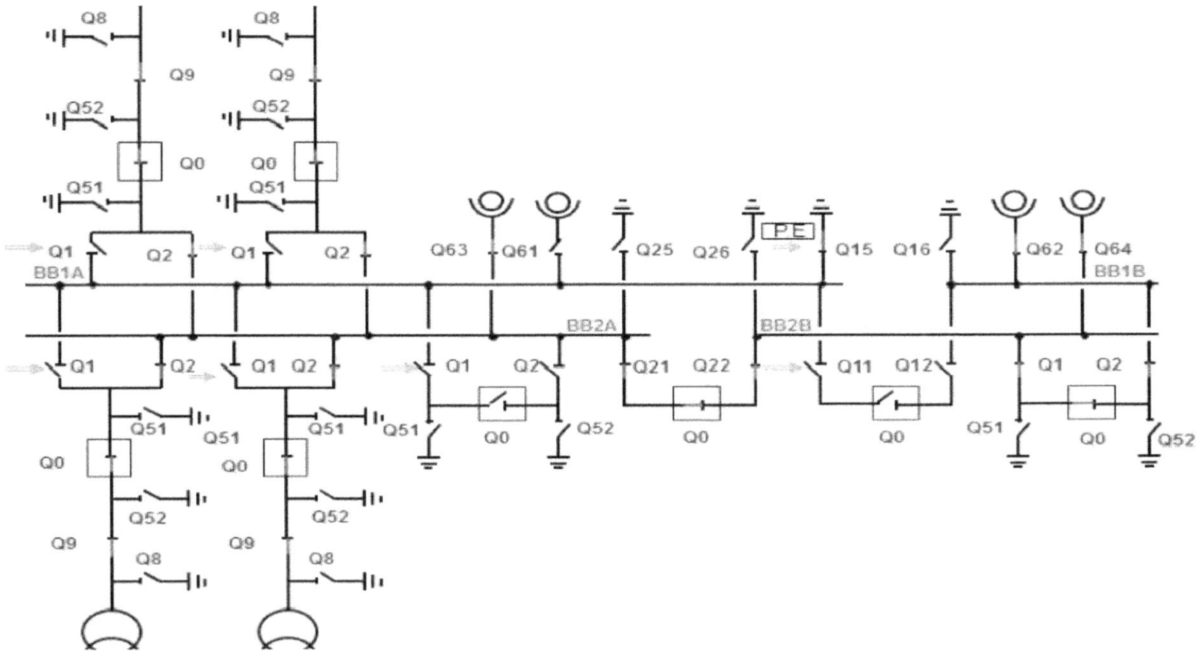

Circuit Breaker Operation

a- Normal Condition:

- Current transformer ratio is 100:5 A. Pick up current = 10A
- At normal condition, the secondary current flowing through the relay is less than the pick up current I_{PU}.
- The relay is energized but not enough to close the contact.
- I_{NORM} = 100A, I_{SEC} = 100 / 20 = 5A less than I_{PU}.
- The main contact of the CB remains closed.

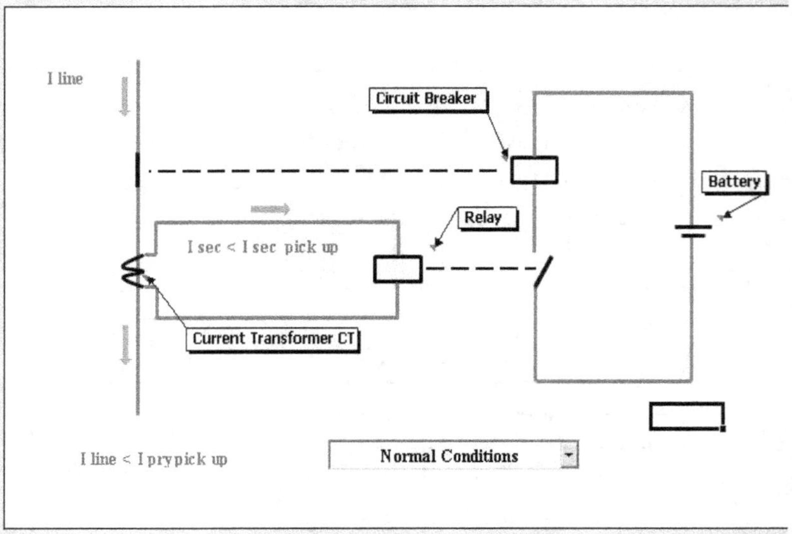

Fig 5. Normal Condition

Circuit Breaker Operation

b- Fault Current :

- When there is a fault current, the secondary current will be greater than the pick up current.
- I_{FAULT} = 1kA, I_{SEC} = 1000/20 = 50A more than I_{PU} = 10A, the relay coil gets enough current to open its contact.

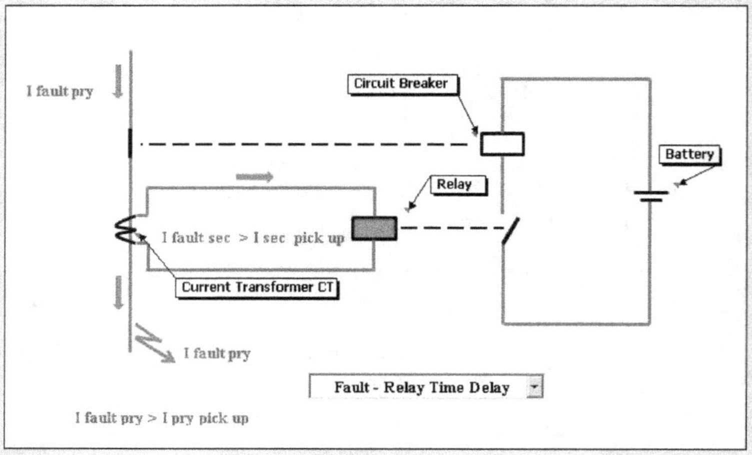

Fig 6. Fault Condition

Circuit Breaker Operation

c- Triggering Circuit Breaker

- The relay contact closes, the circuit breaker coil gets energized from the battery.

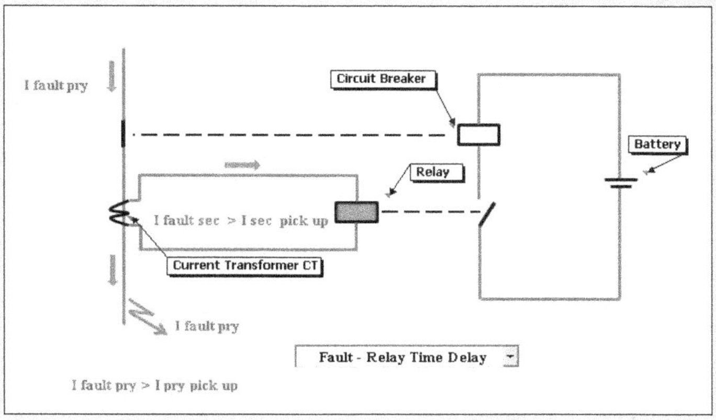

Fig 7. Triggering Condition

Circuit Breaker Operation

d- Clearing Fault

- circuit breaker coil opens the main contact, the fault is now cleared.
- When the fault is isolated, the circuit breaker contact gets closed manually or electrically to restore the power to the system.

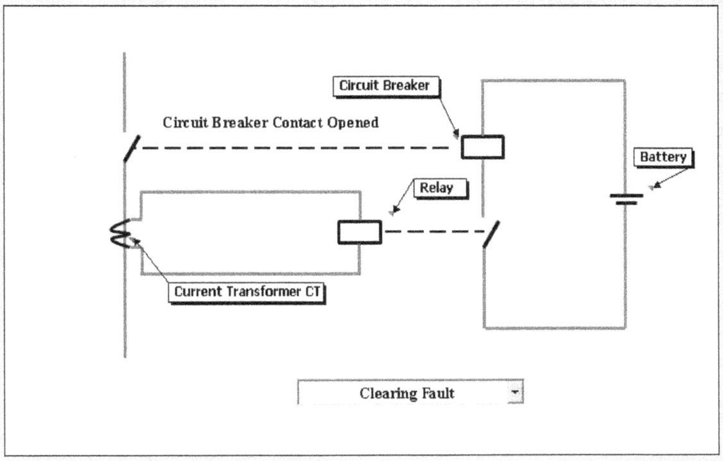

Fig 8. Clearing Fault

LO3

Outline standard methods for power distribution to consumer installations.

LO3-1

Single Line Diagram

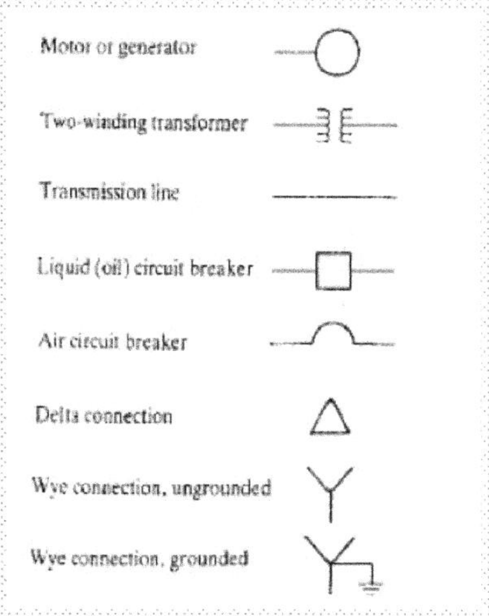

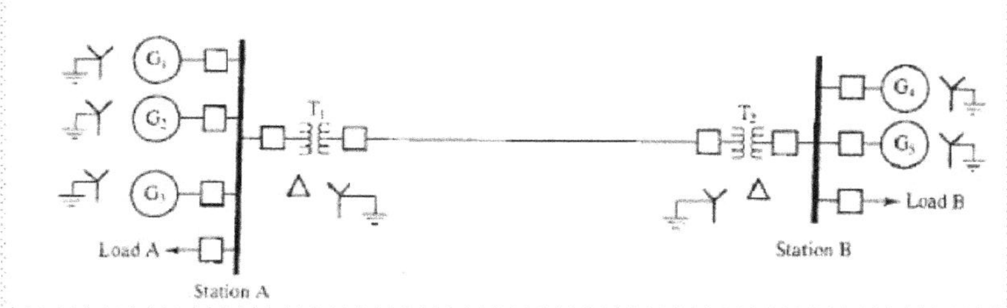

Single Line Diagram

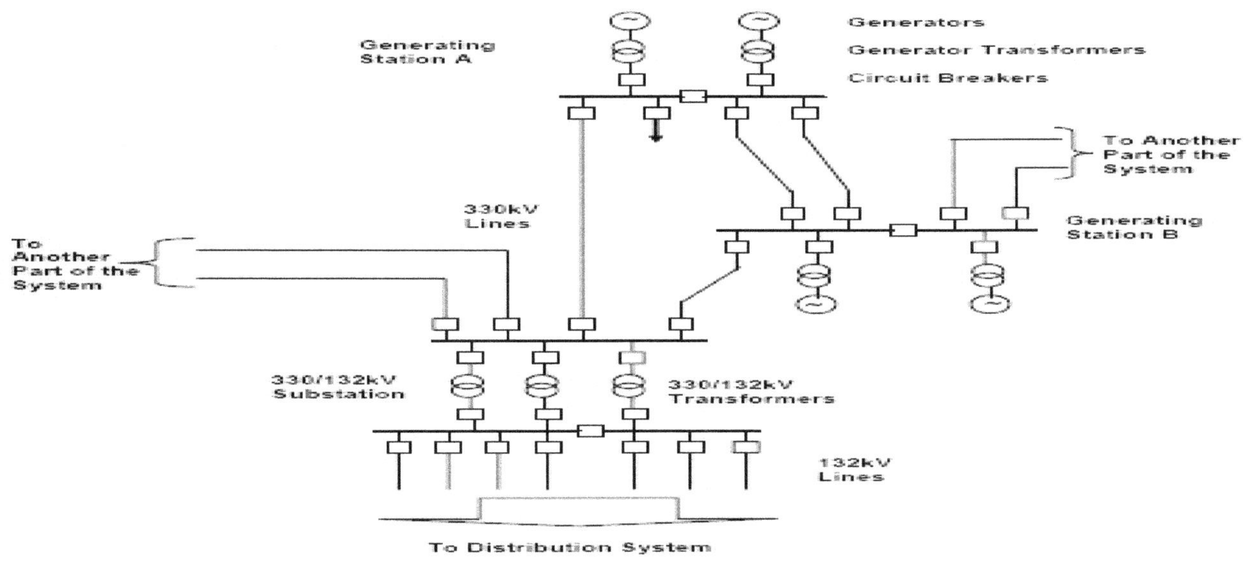

Figure 11. A single-line diagram of a part of a power system

Single Line Diagram

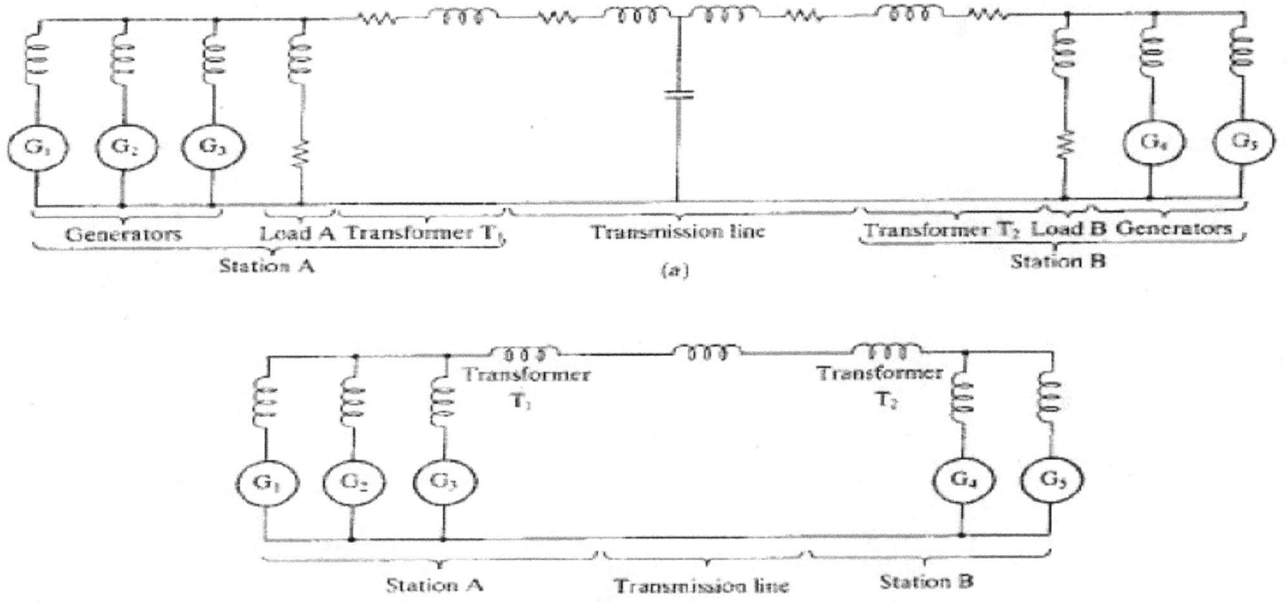

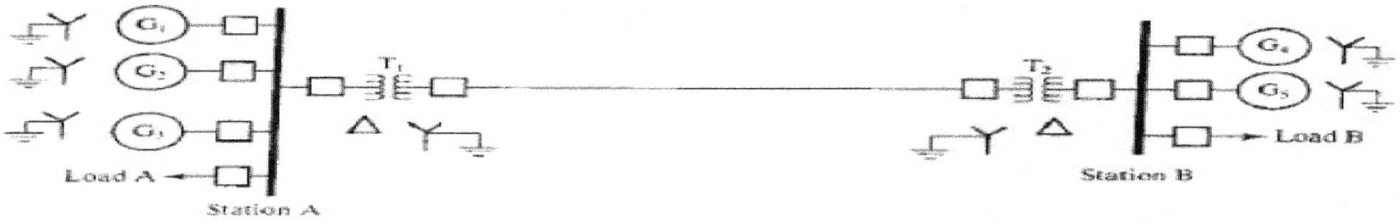

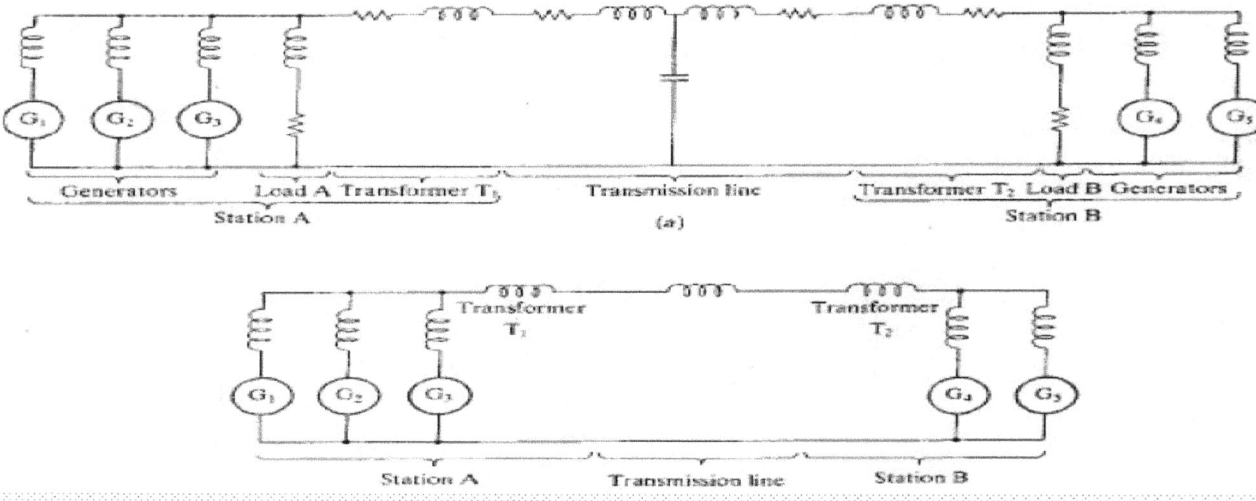

Single Line Diagram

1. A generator can be represented by a voltage source in series with an inductive reactance. The internal resistance of the generator is negligible compared to the reactance.
2. The loads are inductive.
3. The transformer core is ideal, and the transformer may be represented by a reactance.
4. The transmission line is a medium-length line and can be denoted by a T circuit. An alternative representation, such as a π circuit, is equally applicable.
5. The delta-wye-connected transformer T_1 may be replaced by an equivalent wye-wye-connected transformer (via a delta-to-wye transformation) so that the impedance diagram may be drawn on a per-phase basis.

Electrical Shock

Shock protection

- **For shock protection it is always ensured that the *Earth Conductors* are properly connected.**
- **Relatively small currents may be sufficient to kill or injure.**
- **Shocks occur when electric currents flow through the body between points at different voltages.**
- **In summary, the risk and severity of injury depends on two factors:**
  1. **the duration of the shock**
  2. **the amount of current that flows in the body tissues**

Electrical Shock

Disconnection times

- **In short-circuits the current flow could be enormous (thousands of amps).**
- **It is limited only by the resistance of the cable between your house and the supply system, which is usually less than an ohm.**
- **In this kind of fault, the MCB or fuse will trip in its shortest possible time: usually about 0.1 seconds for a fuse and 0.01 seconds for an MCB.**

Types of shock

Electric shocks are of two types:

1. direct contact

2. indirect contact

Direct contact

➢ 'Direct contact' occurs when a body part touches a live part directly. This type of shock is particularly dangerous, as the full voltage of the supply can be developed across the body. In a well-designed electrical installation there should be little or no risk of direct contact; in most cases it arises out of carelessness (e.g. changing a light bulb with the outlet switched on). However, it can sometimes arise from wear and tear, such as the breakdown of insulation on a flexible cable that is badly stressed.

Indirect contact

➢ Indirect contact occurs when a live part touches a piece of metal, and the body comes into contact with the live metal. Indirect contact can occur as a result of faults in electrical appliances, particular with metal casings.

➢ Your main protection against indirect contact is earthing, combined with an overcurrent cut-out device. This works because the large current that will flow to earth in the event of a fault should activate the overcurrent device.

Types of Primary Earth Connections

➢ Your house will (or at least should) contain, in or near the main distribution board, a primary earth terminal. This is the main point to which all circuit earths will be run back. Of course, there may be *other* paths to earth for current elsewhere in the premises. It stands to reason that the main earth terminal should provide a *very* low resistance path to true earth.

➢ There are three main ways that this earth terminal may be connected to a true earth. These are identified by the abbreviations shown in table 2.1.

System Earthing

System earthing is primarily required to ensure that the potential of each conductor of the system restricts to its desired value for which the insulation is provided.The system earthing is so designed that it should facilitate the efficient and quick operation of protective relays in case of any earth fault.

BS 7671 lists five types of earthing system:

TN-S, TN-C-S, TT, TN-C, and IT.

T = Earth (from the French word Terre)

N = Neutral

S = Separate

C = Combined

I = Isolated (The source of an IT system is either connected to earth through a deliberately introduced earthing impedance or is isolated from Earth. All exposed-conductive-parts of an installation are connected to an earth electrode.)

When designing an electrical installation, one of the first things to determine is the type of earthing system. The distributor will be able to provide this information.

Types of Primary Earth Connections

Supply type code	*Meaning*
TN-S	Supplier provides a separate earth connection, usually direct from the distribution station and via the metal sheath of the supply cable
TN-C-S	Supplier provides a combined earth/neutral connection; your main earth terminal is connected to their neutral
TT	Supplier provides no earth; you have an earth spike near your premises

1.1 TN-S system earthing

A TN-S system, shown in fig 1, has the neutral of the source of energy connected with earth at one point only, at or as near as is reasonably practicable to the source, and the consumer's earthing terminal is typically connected to the metallic sheath or armour of the distributor's service cable into the premises.

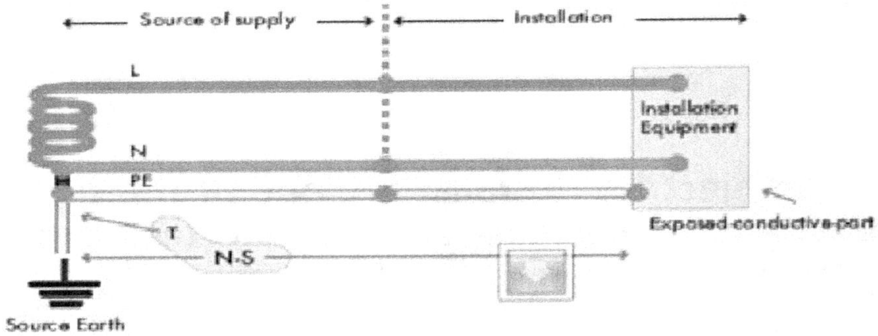

TN-S System

- This is probably the most usual earthing system in the UK, with the Electricity Supply Company providing an earth terminal at the incoming mains position
  - This earth terminal is connected by the supply protective conductor (PE) back to the star point (neutral) of the secondar winding of the supply transformer, which is also connected at that point to an earth electrode
  - The earth conductor usually takes the form of the armour and sheath (if applicable) of the underground supply cable

TN-S System

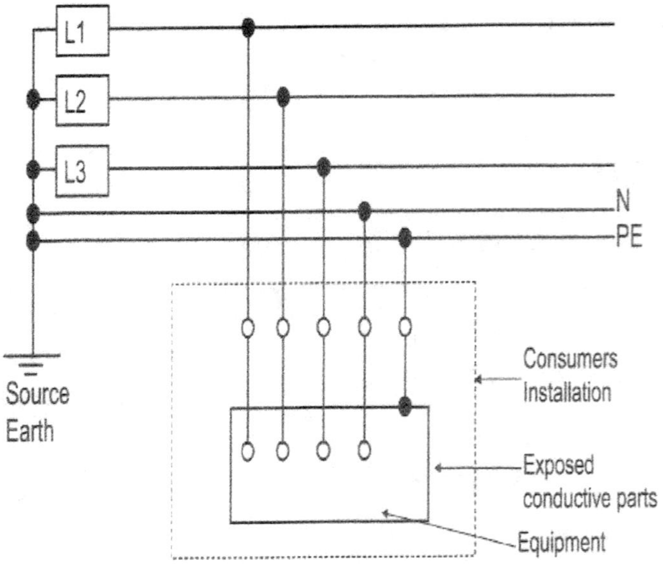

TN-C-S System

- In this system, the installation is TN-S, with separate neutral and protective conductors
  - The supply, however, uses a common conductor for both the neutral and the earth
  - This combined earth and neutral system is sometimes called the 'protective and neutral conductor' (PEN) or the 'combined neutral and earth' conductor (CNE)
  - The system is usually called the protective multiple earth (PME) system

TN-C-S System

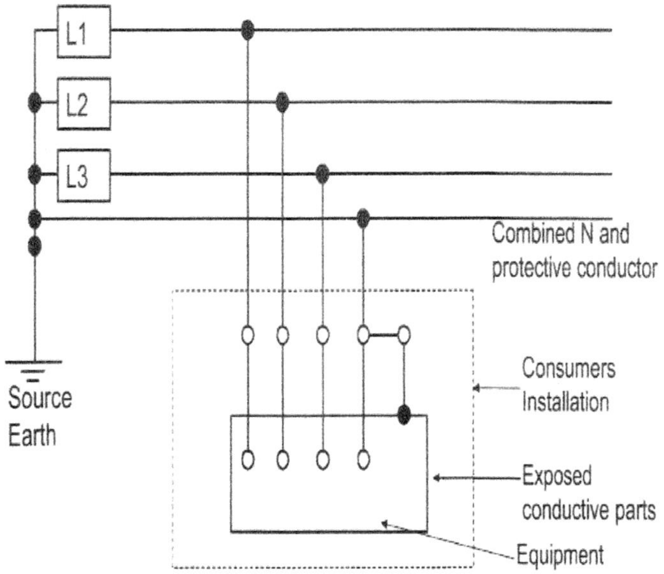

1.2 TN-C-S system earthing

A TN-C-S system, shown in fig 3, has the supply neutral conductor of a distribution main connected with earth at source and at intervals along its run. This is usually referred to as protective multiple earthing (PME). With this arrangement the distributor's neutral conductor is also used to return earth fault currents arising in the consumer's installation safely to the source. To achieve this, the distributor will provide a consumer's earthing terminal which is linked to the incoming neutral conductor.

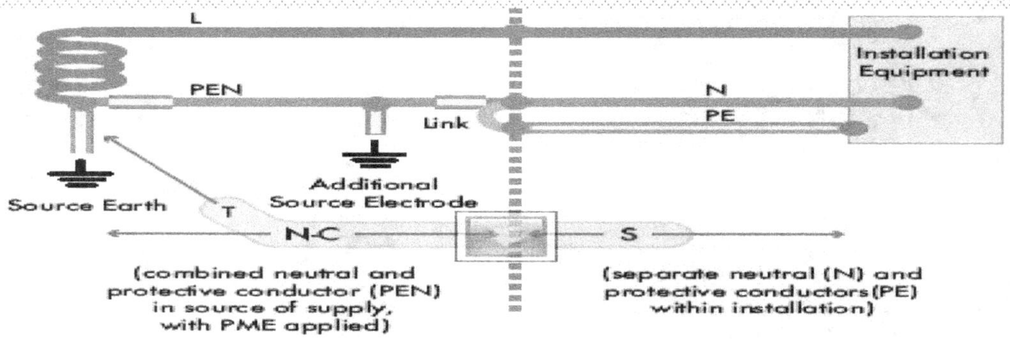

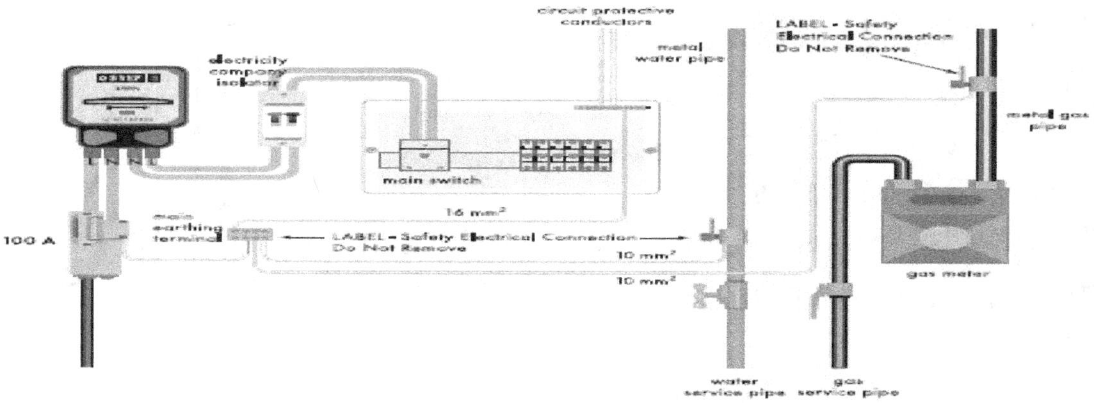

- T T System of Earthing

In T T System of earthing, the source is earthed. But the conductive parts of the installations are connected to the earth through one or more local earth electrodes. These local electrodes does not have any direct connection to the earthing system of source. This T T system of earthing is applicable for both three phase and single phase installations.

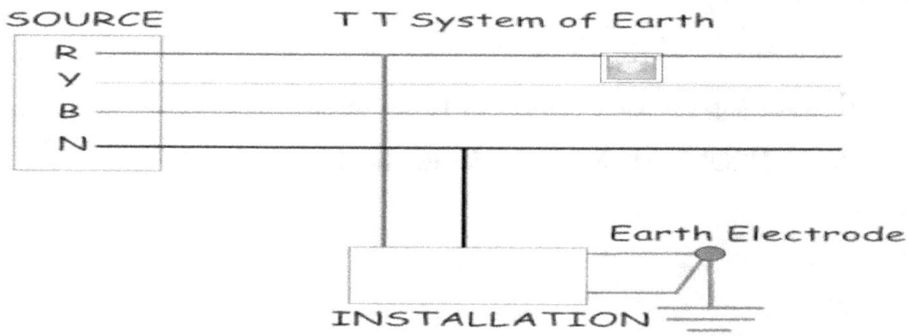

1.3 TT system earthing

A TT system, shown above, has the neutral of the source of energy connected as for TN-S, but no facility is provided by the distributor for the consumer's earthing. With TT, the consumer must provide their own connection to earth, i.e. by installing a suitable earth electrode local to the installation.

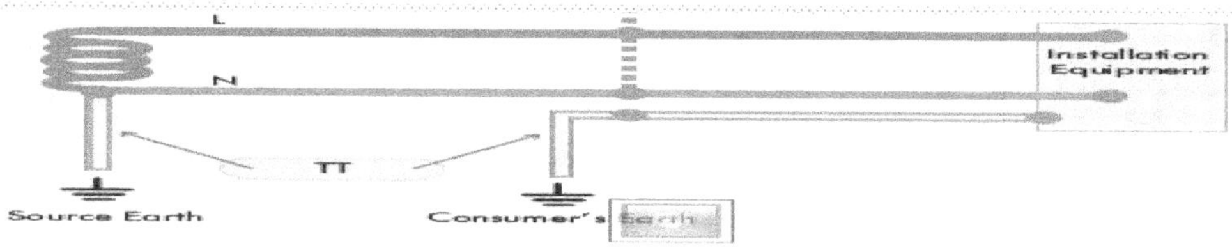

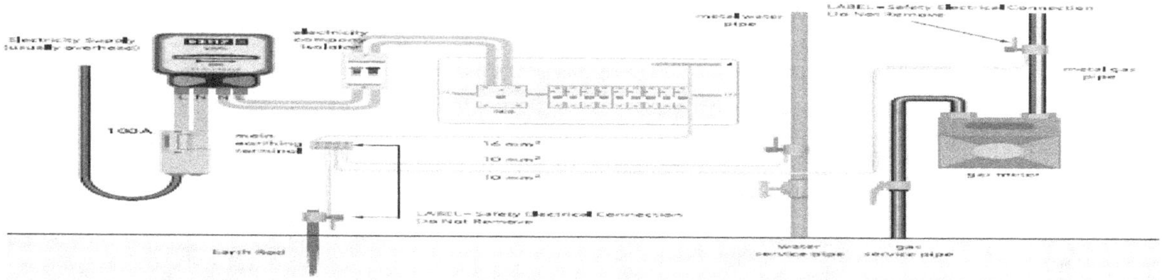

TT Systems

- This arrangement covers installations not provided with an earth terminal by the Electricity Supply Company
  - Thus it is the method employed by most (usually rural) installations fed by an overhead supply
  - Neutral and earth (protective) conductors must be kept quite separate throughout the installation, with the final earth terminal connected to an earth electrode by means of an earthing conductor
- Effective earth connection is sometimes difficult
  - Because of this, socket outlet circuits must be protected by a residual current device (RCD) with an operating current of 30 mA

TT System

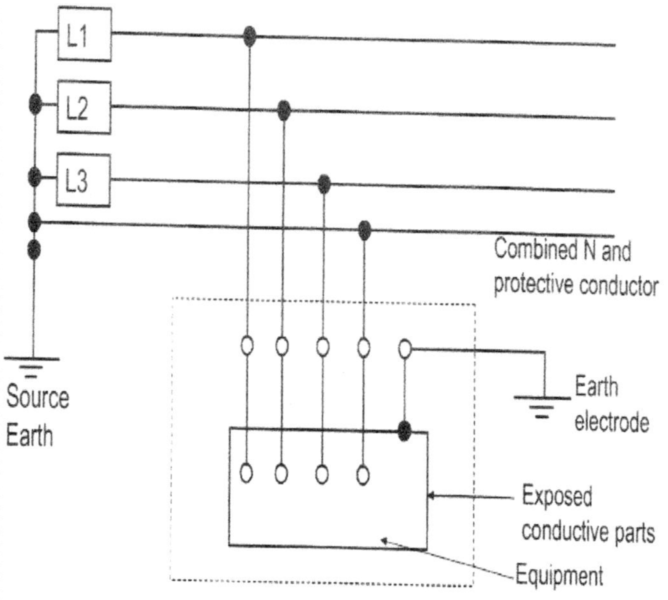

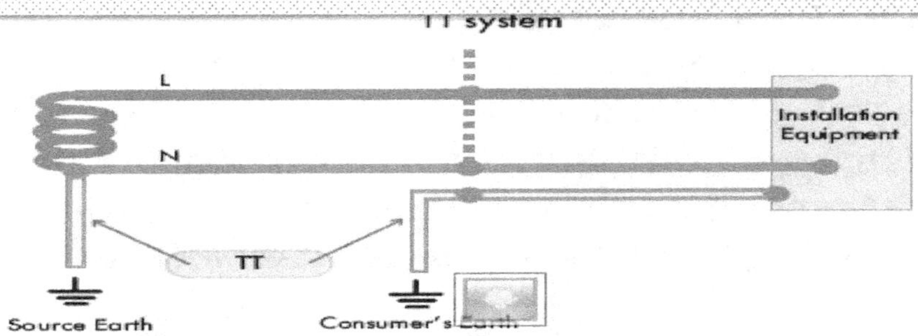

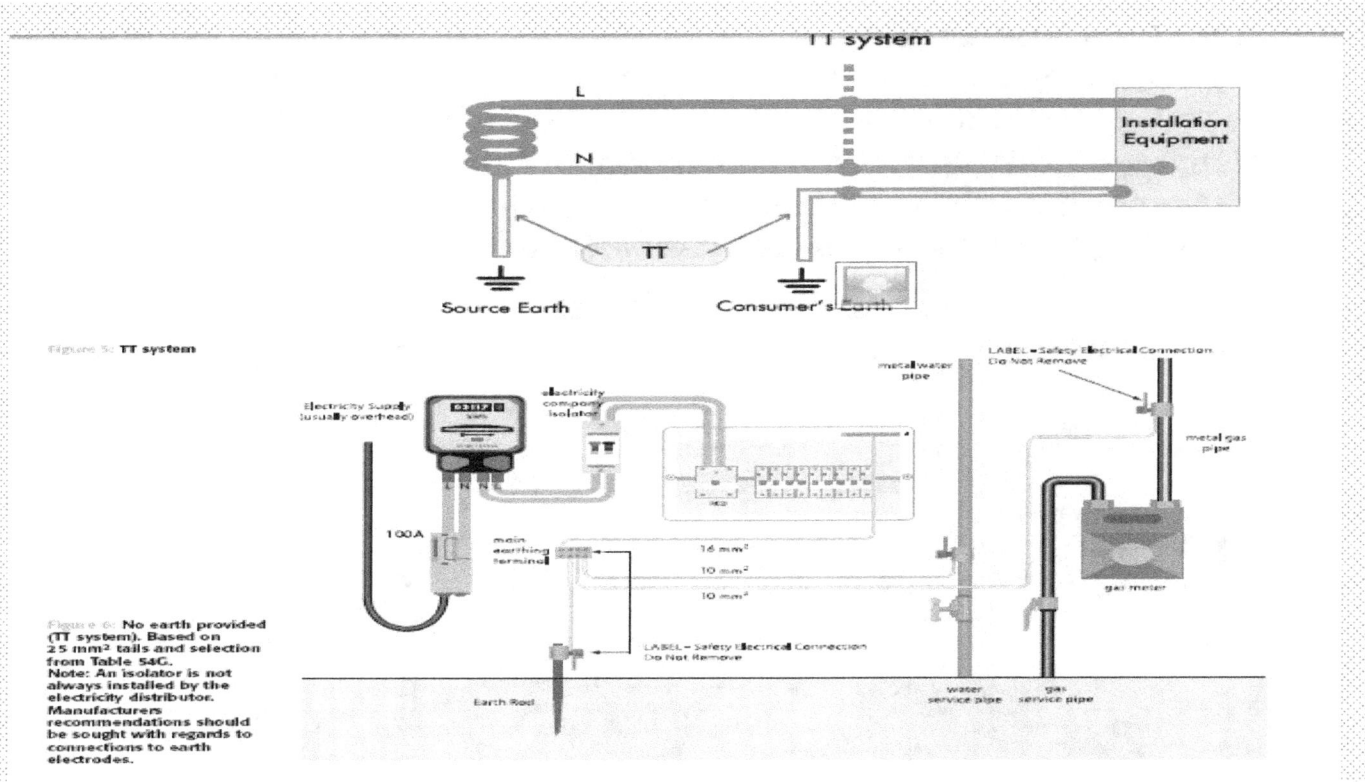

Figure 5: TT system

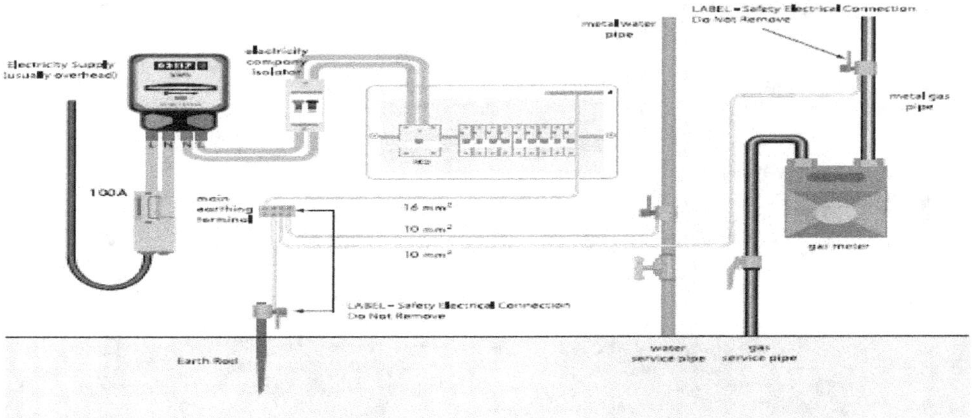

Figure 6: No earth provided (TT system). Based on 25 mm² tails and selection from Table 54G.
Note: An isolator is not always installed by the electricity distributor. Manufacturers recommendations should be sought with regards to connections to earth electrodes.

- I T System of Earthing

I T System of earthing is generally used in unearthed three phase network. Here, the three phase source is isolated from earth or connected to earth through a high impedance of suitable value. The conductive parts including metal body of the installations are connected to the earth through one or more local earth electrodes. These local electrodes does not have any direct connection to the source.

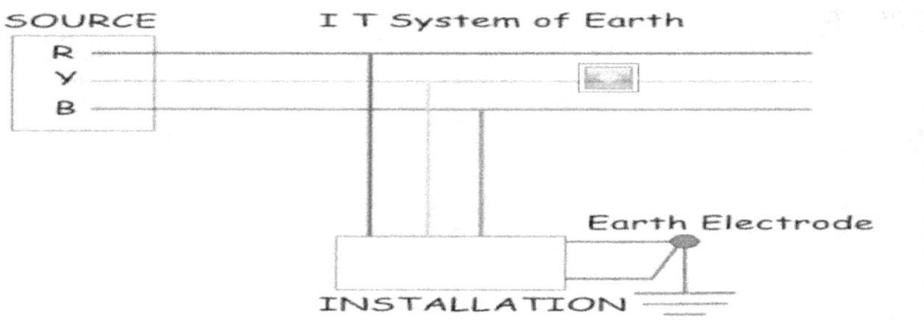

IT System

- The installation arrangements in the IT system are the same for those of the TT system
  - However, the supply earthing is totally different
  - The IT system can have an unearthed supply, or one which is not solidly earthed but is connected to earth through a current limiting impedance
  - The total lack of earth in some cases, or the introduction of current limiting into the earth path, means that the usual methods of protection will not be effective
  - For this reason, IT systems are not allowed in the public supply system in the UK (- an exception is in medical situations such as hospitals)
  - The method is also sometimes used where a supply for special purposes is taken from a private generator

IT System

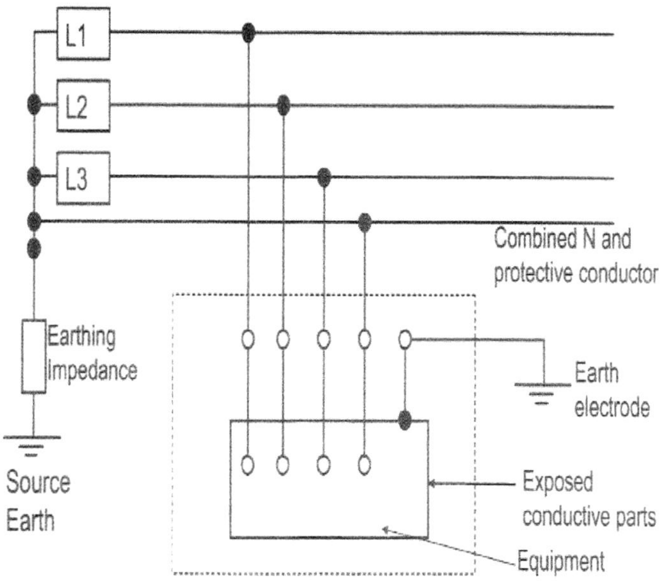

Earthing System Classification

- The second letter indicates the earthing arrangement in the installation
  - **T** - all exposed conductive metalwork is connected directly to earth
  - **N** - all exposed conductive metalwork is connected directly to an earthed supply conductor provided by the Electricity Supply Company
- The third and fourth letters indicate the arrangement of the earthed supply conductor system
  - **S** - neutral and earth conductor systems are quite separate
    C - neutral and earth are combined into a single conductor
- A number of possible combinations of earthing systems in common use are indicated in the following subsections

Earthling Loop

➢ If a live part in the appliance comes into contact with the casing, a current will flow from the live of the supply company's apparatus to the premises, along the live conductor to the appliance, through the fault to the casing, from the casing to the earth terminal via the earth conductor, and from the earth terminal back to the supply company's apparatus via its earth connection.

➢ This circuit is called the `earth loop'. Note that part of the earth loop is outside your premises, and in the supply company's cables and apparatus. You have no control over that part of the loop. The part of the loop inside your house has a resistance which can be calculated, because we know the resistances per meter of the various cables that are likely to be used.

➢ If an earth fault occurs (that is, a short-circuit between live and earth), the path for current includes that supply company's live conductor into your premises, the live part of your cabling, the earth part of your cabling, and the earth part of the supplier's system. You can calculate the resistance of your part this system, or measure it, but you may need to approach the supply company for the resistance of their part. Suppliers are legally obliged to tell you this; it is, after all, very important for ensuring safety.

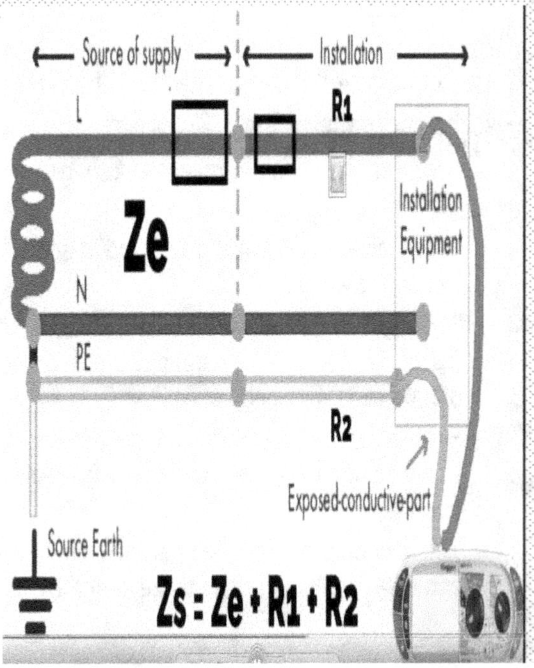

➢ When a short-circuit from live to earth occurs, the earth loop is the path that the current will flow in. This current could be very large; it should certainly be large enough to blow the fuse or trip the MCB before serious injury occurs. This suggests that the earth loop resistance should be as low as possible.

➢ For initial and approximate design calculations, you can use the `worst case' values of earth loop resistance given in table A.11. If your installation appears safe with these worst case figures, it will almost certainly prove to be safe with the true figures. However, the earth loop resistance figures depend on your knowing the supply type of your premises. If you don't know this, you will need to ask the supply company anyway.

LO3

Outline standard methods for power distribution to consumer installations.

LO3-2

Cable Size

with previous editions of the IET Regulations, but it has not been so clearly indicated.

How then do we begin to design? Clearly, plunging into calculations of cable size is of little value unless the type of cable and its method of installation is known. This in turn will depend on the installation's environment. At the same time, we would need to know whether the supply was single or three phase, the type of earthing arrangements and so on. Here

Basically there are eight stages in such a procedure. These are the same whatever the type of installation, be it a cooker circuit or a sub-main cable feeding a distribution board in a factory. Here then are the eight basic steps in a simplified form:

1. Determine the design current I_b.
2. Select the rating of the protection I_n.
3. Select the relevant rating factors.
4. Divide I_n by the relevant rating factors (CFs) to give tabulated cable current-carrying capacity I_t.
5. Choose a cable size to suit I_t.
6. Check the voltage drop.
7. Check for shock risk constraints.
8. Check for thermal constraints.

1. Determining Design Current Ib

Design current

In many instances the design current I_b is quoted by the manufacturer, but there are times when it has to be calculated. In that case there are two formulae involved, one for single phase and one for three phase:

Single phase:

$$I_b = \frac{P}{V} \quad (V \text{ usually } 230\,V)$$

Three phase:

$$I_b = \frac{P}{\sqrt{3} \times V_L} \quad (V_L \text{ usually } 400\,V)$$

If an item of equipment has a power factor (PF) and/or has moving parts, efficiency (eff) will have to be taken into account. Hence:

Single phase:

$$I_b = \frac{P}{V \times PF \times eff}$$

Three phase:

$$I_b = \frac{P}{\sqrt{3} \times V_L \times PF \times eff}$$

2. Determining rating of protection In

Nominal setting of protection

Having determined I_b we must now select the nominal setting of the protection I_n such that $I_n \geq I_b$. This value may be taken from IET Regulations, Tables 41.2, 41.3 or 41.4, or from manufacturers' charts. The choice of fuse or circuit breaker (cb) type is also important and may have to be changed if cable sizes or loop impedances are too high.

Example

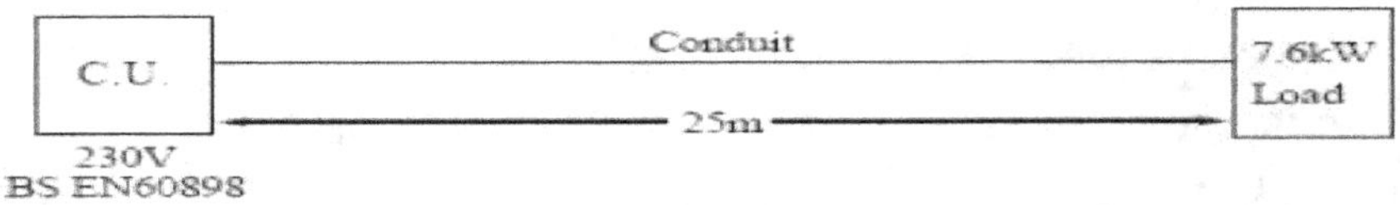

Design current i/. $I_b = \dfrac{P}{U_o}$

$$I_b = \frac{7600}{230} = \underline{33.04A}$$

Rating of the protective device ii/. $I_n = \underline{40A}$ (next size up)

3- *Installation Reference Method*

Table A6(1) : Method 3

TABLE A6(1)

Single-core PVC insulated cables, non-armoured, with or without sheath

(COPPER CONDUCTORS)

BS 6004
BS 6231
BS 6346

CURRENT CARRYING CAPACITY (Amperes)

Ambient temperature: 30°C
Conductor operating temperature: 70°C

Conductor cross-sectional area	Reference Method 4 (enclosed in conduit in thermally insulating wall etc.)		Reference Method 3 (enclosed in conduit on a wall or in trunking etc.)		Reference Method 1 (clipped direct)		Reference Method 11 (on a perforated cable tray horizontal or vertical)		Reference Method 12 (free air) Horizontal flat spaced	Reference Method 12 (free air) Vertical flat spaced	Reference Method 12 (free air) Trefoil
	2 cables, single-phase a.c. or d.c.	3 or 4 cables three-phase a.c.	2 cables, single-phase a.c. or d.c.	3 or 4 cables three-phase a.c.	2 cables, single-phase a.c. or d.c. flat and touching	3 or 4 cables three-phase a.c. flat and touching or trefoil	2 cables, single-phase a.c. or d.c. flat and touching	3 or 4 cables three-phase a.c. flat and touching or trefoil	2 cables, single-phase a.c. or d.c. or 3 cables three-phase a.c.	2 cables, single-phase a.c. or d.c. or 3 cables three-phase a.c.	3 cables trefoil three phase a.c.
1	2	3	4	5	6	7	8	9	10	11	12
mm²	A	A	A	A	A	A	A	A	A	A	A
1	11	10.5	13.5	12	15.5	14	—	—	—	—	—
1.5	14.5	13.5	17.5	15.5	20	18	—	—	—	—	—
2.5	20	18	24	21	27	25	—	—	—	—	—
4	26	24	32	28	37	33	—	—	—	—	—
6	34	31	41	36	47	43	—	—	—	—	—
10	46	42	57	50	65	59	—	—	—	—	—
16	61	56	76	68	87	79	—	—	—	—	—
25	80	73	101	89	114	104	131	114	146	130	110
35	99	89	125	119	141	129	162	143	181	162	137
50	119	108	151	134	182	167	196	174	219	197	167
70	151	136	192	171	234	214	251	225	281	254	216
95	182	164	232	207	284	261	304	275	341	311	264
120	210	188	269	239	330	303	352	321	396	362	308
150	240	216	300	262	381	349	406	372	456	419	356
185	273	245	341	296	436	400	463	427	521	480	409
240	321	286	400	346	515	472	546	507	615	569	485
300	367	328	458	394	594	545	629	587	709	659	561
400	—	—	546	467	694	634	754	689	852	795	656
500	—	—	626	533	792	723	868	789	982	920	749
630	—	—	720	611	904	826	1 005	905	1 138	1 070	855
800	—	—	—	—	1 030	943	1 086	1 020	1 265	1 188	971
1000	—	—	—	—	1 154	1 058	1 216	1 149	1 420	1 317	1 079

Conduits

Trucking

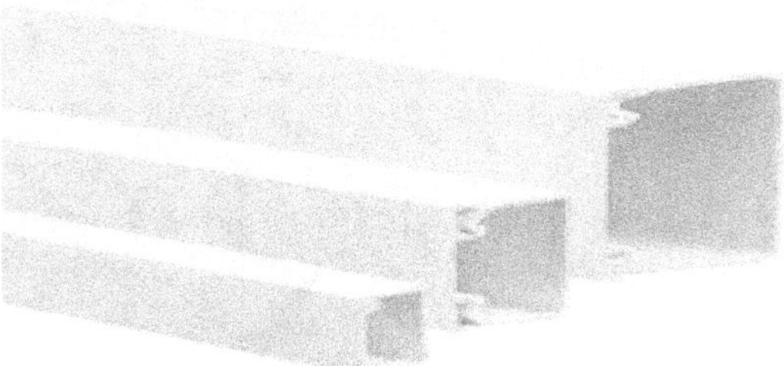

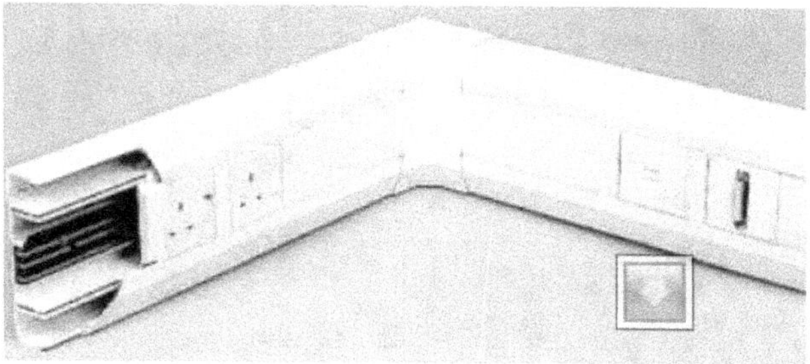

3. Select the Relevant Rating Factors

Rating factors

When a cable carries its full-load current it can become warm. This is not a problem unless its temperature rises further due to other influences, in which case overheating could damage the insulation. These other influences are: high ambient temperature; cables grouped together closely; uncleared overcurrents; and contact with thermal insulation.

For each of these conditions there is a rating factor which will respectively be called C_a, C_g, C_i and C_i respectively, and which derates cable current-carrying capacity or conversely increases cable size (IET

Ambient temperature – **The standard or reference temperature for the installation of cables is 30_{o}C.**

- **If cables are installed in conditions where the ambient temperature is in excess of this then a correction factor must be able when calculating the required cable cross section.**
- **A table of correction factors for ambient temperature may be viewed.**
- **The symbol for the Correction factor for ambient temperature is Ca.**

Ambient temperature C_a

The cable ratings in the IET Regulations are based on an ambient temperature of 30°C, and hence it is only above this temperature that an adverse correction is needed. Table 4B1 of the Regulations gives factors for all types of protection.

- Rubber Insulated Cables

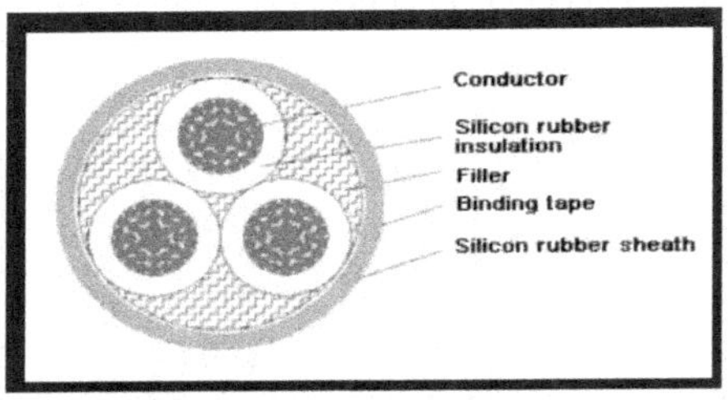

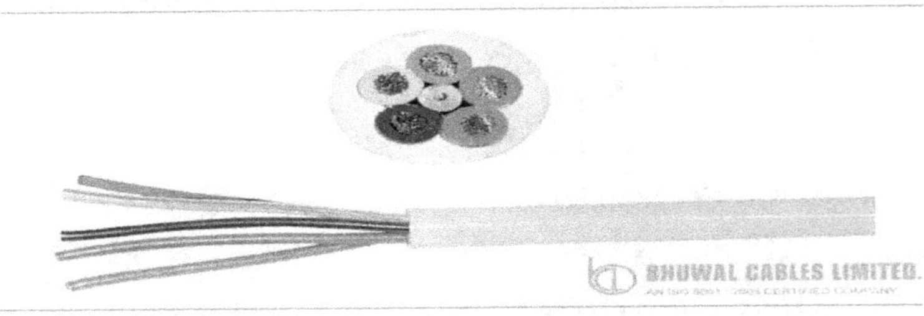

- PVC Cables

2.5 sqmm x 3 core pvc insulation copper wire cable

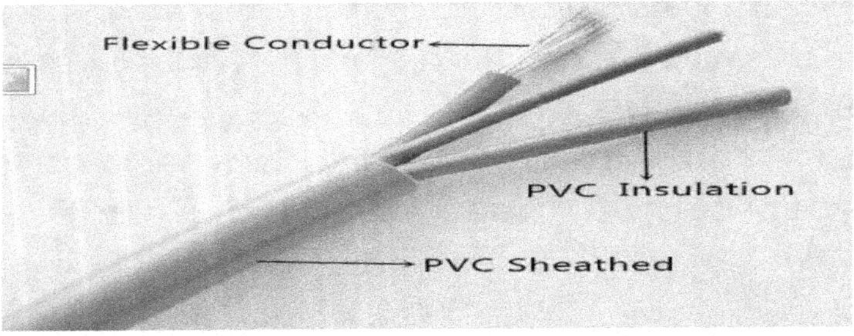

Table A5(1)

Correction factors for ambient temperature

Note: This table applies where the associated overcurrent protective device is intended to provide short circuit protection only. Except where the device is a semi-enclosed fuse to BS3036 the table also applies where the device is intended to provide overload protection.

Type of insulation	Operating temperature	Ambient Temperature °C														
		25	30	35	40	45	50	55	60	65	70	75	80	85	90	95
Rubber (flexible cables only)	60°C	1.04	1.0	0.91	0.82	0.71	0.58	0.41	—	—	—	—	—	—	—	—
General purpose PVC	70°C	1.03	1.0	0.94	0.87	0.79	0.71	0.61	0.50	0.35	—	—	—	—	—	—
Paper	80°C	1.02	1.0	0.95	0.89	0.84	0.77	0.71	0.63	0.55	0.45	0.32	—	—	—	—
Rubber	85°C	1.02	1.0	0.95	0.90	0.85	0.80	0.74	0.67	0.60	0.52	0.43	0.30	—	—	—

Table 4.3 Correction factors to current rating for ambient temperature (Ca) (from [Tables 4C1 and 4C2] of BS 7671: 1992)

Ambient temperature	Type of insulation			
(°C)	70°C p.v.c	85°C rubber	70°C m.i	105°C m.i
25	1.03 (1.03)	1.02 (1.02)	1.03 (1.03)	1.02 (1.02)
30	1.00 (1.00)	1.00 (1.00)	1.00 (1.00)	1.00 (1.00)
35	0.94 (0.97)	0.95 (0.97)	0.93 (0.96)	0.96 (0.98)
40	0.87 (0.94)	0.90 (0.95)	0.85 (0.93)	0.92 (0.96)
45	0.79 (0.91)	0.85 (0.93)	0.77 (0.89)	0.88 (0.93)
50	0.71 (0.97)	0.80 (0.91)	0.67 (0.86)	0.84 (0.91)
55	0.61 (0.84)	0.74 (0.88)	0.57 (0.79)	0.80 (0.89)

Figures in brackets apply to semi-enclosed fuses used for overload protection

Grouping – Where cables are run in contact with each other heat dissipation is made more difficult as cables can impose an external heating effect on each other.

- **This must again be accounted for in determining the current carrying capacity for a cable.**
- **Table 4C1 of BS7671 gives values for cables that are grouped and touching.**
- **The symbol for correction factor for cable grouping is Cg**

Grouping C_g

When cables are grouped together they impart heat to each other. Therefore the more cables there are, the more heat they will generate, thus increasing the temperature of each cable. Table 4C1 of the Regulations gives factors for such groups of cables or circuits. It should be noted that the figures given are for cables of the same size, and hence **no** correction need to be made for cables grouped at the outlet of a domestic consumer unit, for example where there is a mixture of different sizes.

A typical situation where rating factors need to be applied would be in the calculation of cable sizes for a lighting system in a large factory. Here many cables of the same size and loading may be grouped together in trunking and could be expected to be fully loaded all at the same time.

- Because of this, cables installed in groups with others (for example, if enclosed in a conduit or trunking) are allowed to carry less current than similar cables clipped to, or lying on, a solid surface which can dissipate heat more easily. If surface mounted cables are touching the reduction in the current rating is, as would be expected, greater than if they are separated.
- For example, if a certain cable has a basic current rating of 24 A and is installed in a trunking with six other circuits (note carefully, this is circuits and not cables), Cg has a value of 0.57 and the cable current rating becomes 24 / 0.57 A. The symbol Cg is used to represent the factor used for derating cables to allow for grouping. {Table 4.4} shows some of the more useful values of Cg.
-

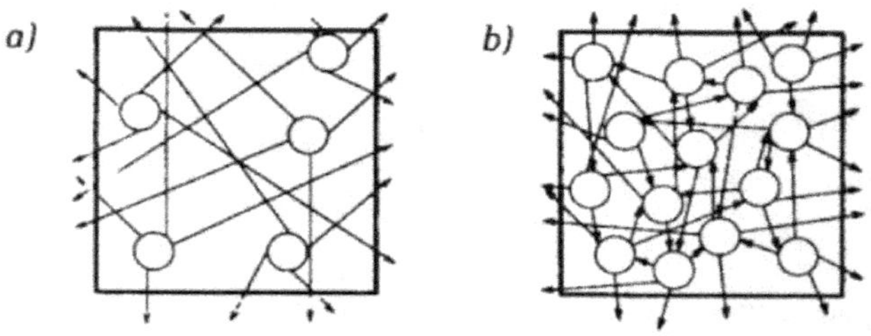

Fig 4.9 The need for the grouping correction factor Cg

a) widely spaced cables dissipate heat easily

b) A closely packed cable cannot easily dissipate heat and so its temperature rises

Table 4.4 - Correction factors for groups of cables

Number of circuits	Correction factor Cg		
-	Enclosed or clipped	Clipped to non-metallic surface	
-	-	Touching	Spaced*
2	0.80	0.85	0.94
3	0.70	0.79	0.90
4	0.65	0.75	0.90
5	0.60	0.73	0.90
6	0.57	0.72	0.90
7	0.54	0.72	0.90
8	0.52	0.71	0.90
9	0.50	0.70	0.90

- Clipped Cable

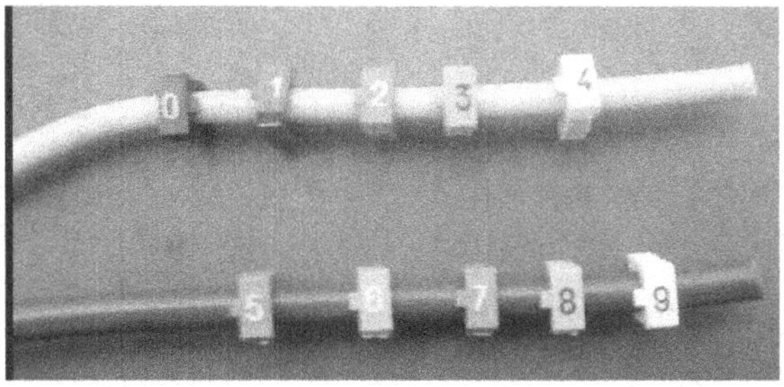

- Enclosed Cable

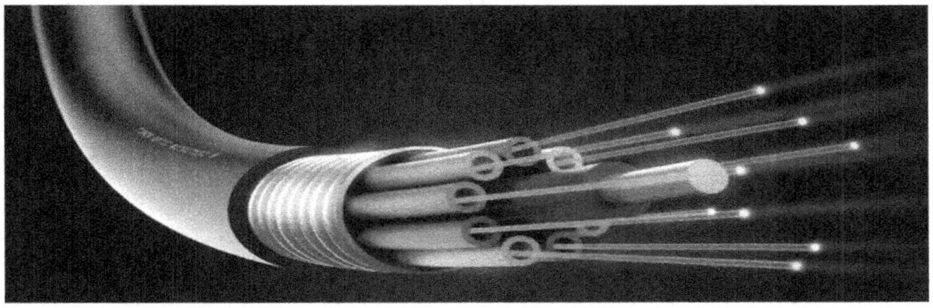

Thermal Insulation – **If a cable, through any part of its run, passes completely through thermal insulation this can greatly impede its ability to dissipate heat generated through current flow. Therefore regulation 523.7 identifies corrections to be made for this situation. The symbol for correction factor for thermal insulation is Ci.**

Thermal insulation C_i

With the modern trend towards energy saving and the installation of thermal insulation, there may be a need to derate cables to account for heat retention.

The values of cable current-carrying capacity given in appendix 4 of the IET Regulations have been adjusted for situations when thermal insulation touches one side of a cable. However, if a cable is totally surrounded by thermal insulation for more than 0.5 m, a factor of 0.5 must be applied to the tabulated clipped direct ratings. For less than 0.5 m, derating factors (Table 52.2) should be applied.

- **The use of thermal insulation in buildings, in the forms of cavity wall filling, roof space blanketing, and so on. is now standard. Since the purpose of such materials is to limit the transfer of heat, they will clearly affect the ability of a cable to dissipate the heat build up within it when in contact with them.**

- **The cable rating tables of [Appendix 4] allow for the reduced heat loss for a cable which is enclosed in an insulating wall and is assumed to be in contact with the insulation on one side. In all other cases, the cable should be fixed in a position where it is unlikely to be completely covered by the insulation. Where this is not possible and a cable is buried in thermal insulation for 0.5 m (500 mm) or more, a rating factor (the symbol for the thermal insulation factor is Ci) of 0.5 is applied, which means that the current rating is halved.**

Table 4.5 - Derating factors (CI) for cables up to 10mm² in cross-sectional area buried in thermal insulation.

Length in insulation (mm)	Derating factor (CI)
50	0.89
100	0.81
200	0.68
400	0.55
500 or more	0.50

Protective Device – Where the circuit protective device is a semi-enclosed fuse to BS3036 a correction factor must be applied to make up for potential hazards that may otherwise arise under overload conditions.

- **The symbol for correction factor for a BS3036 fuse is Cc.**
- **it has a value of 0.725.**

Protection by BS 3036 fuse C_f

As we have already discussed in Chapter 17, because of the high fusing factor of BS 3036 fuses, the rating of the fuse I_n should be less than or equal to $0.725 I_z$. Hence 0.725 is the rating factor which is to be used when BS 3036 fuses are used.

4. Divide In by the relevant rating factors (CFs) to give tabulated cable current carrying capacity

Tabulated Current Carry Capacity I_t

$$I_t \geq I_n / (Ca \times Ci \times Cg \times Cc)$$

Cable Size Calculation

To select a cable for a particular application, take the following steps: (note that to save time it may be better first to ensure that the expected cable for the required length of circuit will] not result in the maximum permitted volt drop being exceeded {4.3.11}).

1. - Calculate the expected (design) current in the circuit (Ib)

2. - Choose the type and rating of protective device (fuse or circuit breaker) to be used (In)

3. - Divide the protective device rated current by the ambient temperature correction factor (Ca) if ambient temperature differs from 30°C

4. - Further divide by the grouping correction factor (Cg)

5. - Divide again by the thermal insulation correction factor (CI)

6. - Divide by the semi-enclosed fuse factor of 0.725 where applicable

7. - The result is the rated current of the cable required, which must be chosen

Cable volt drop

- All cables have resistance, and when current flows in them this results in a volt drop. Hence, the voltage at the load is lower than the supply voltage by the amount of this volt drop.

The volt drop may be calculated using the basic Ohm's law formula	
U = **I** x **R**	
where	**U** is the cable volt drop (V
	I is the circuit current (A), and
	R is the circuit resistance W(Ohms)

- [525-01-03] indicates that the voltage at any load must never fall so low as to impair the safe working of that load, or fall below the level indicated by the relevant British Standard where one applies.

- [525-01-02] indicates that these requirements will be met if the voltage drop does not exceed 4% of the declared supply voltage. If the supply is single-phase at the usual level of 240 V, this means a maximum volt drop of 4% of 240 V which is 9.6 V, giving (in simple terms) a load voltage as low as 230.4 V. For a 415 V three-phase system, allowable volt drop will be 16.6 V with a line load voltage as low as 398.4 V.

Each cable rating in the Tables of [Appendix 4] has a corresponding volt drop figure in millivolts per ampere per metre of run (mV/A/m). Strictly this should be mV/(A m), but here we shall follow the pattern adopted by BS 7671: 1992. To calculate the cable volt drop:

1. Take the value from the volt drop table (mV/A/m)
2. Multiply by the actual current in the cable (NOT the current rating)
3. Multiply by the length of run in metres
4. Divide the result by one thousand (to convert millivolts to volts).

Cross sectional area	conduit in thermal insulation	conduit in thermal insulation	In conduit on wall	In conduit on wall	Clipped direct	Clipped direct	Volt drop	Volt drop
(mm²)	(A)	(A)	(A)	(A)	(A)	(A)	(mV/A/m)	(mV/A/m)
-	2 core	3 or 4 core	2 core	3 or 4 core	2 core	3 or 4 core	2 core	3 or 4 core
1.0	11.0	10.0	13.0	11.5	15.0	13.5	44.0	38.0
1.5	14.0	13.0	16.5	15.0	19.5	17.5	29.0	25.0
2.5	18.5	17.5	23.0	20.0	27.0	24.0	18.0	15.0
4.0	25.0	23.0	30.0	27.0	36.0	32.0	11.0	9.5
6.0	32.0	29.0	38.0	34.0	46.0	41.0	7.3	6.4
10.0	43.0	39.0	52.0	46.0	63.0	57.0	4.4	3.8
16.0	57.0	52.0	69.0	62.0	85.0	76.0	2.8	2.4

Cable Color Code

	Single Phase	Three Phase
Phase Conductor (Line)	Red or Yellow or Blue	Line 1 Red Line 2 Yellow Line 3 Blue
Neutral Conductor	Black	
Protective Conductor (Earth)	Green-and-Yellow	

Cable Color Code

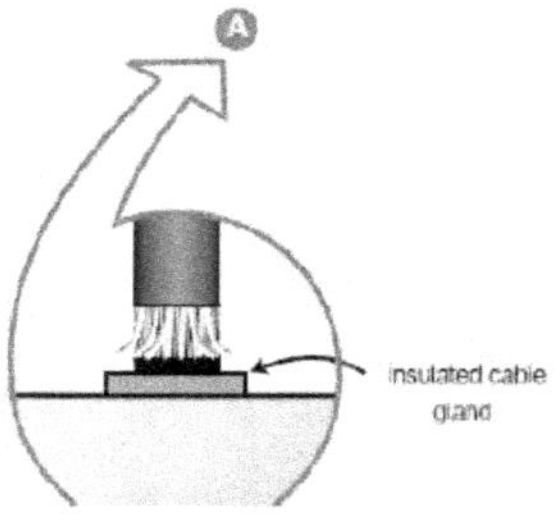

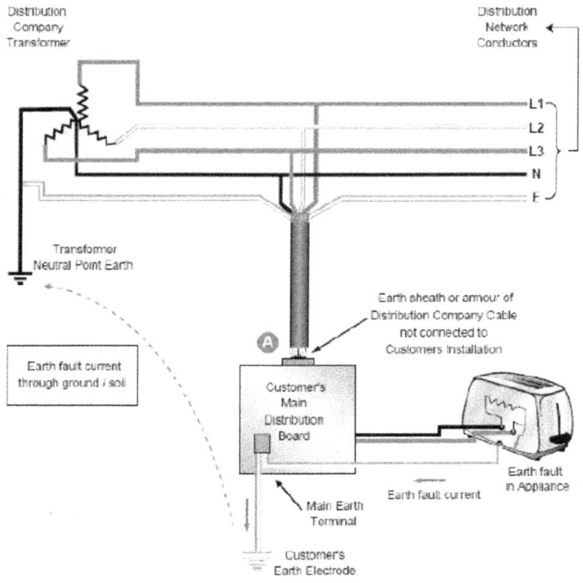

Examples

1- A 7.6kW single-phase load is fed from a distribution board 25m away. The load is fed via PVC singles and the cable is run on its own in steel conduit on the surface, and is protected via a BSEN 60898 Type B circuit breaker. The cable is run in an area where the ambient temperature is 35°C. Determine the size of cable required.

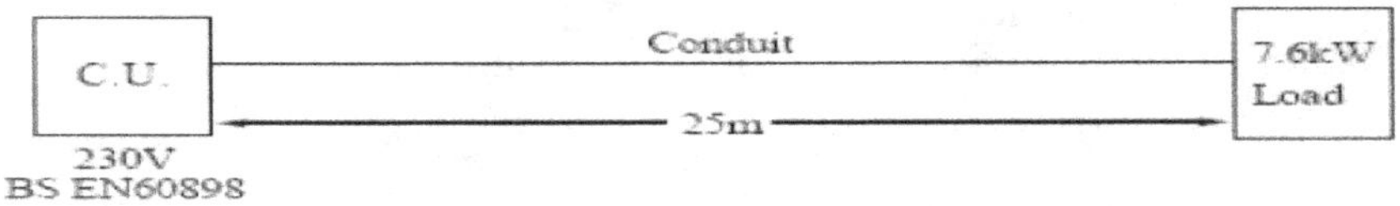

1- ***Grouping***

No factor applies

2- ***Ambient temperature***

Table A5(1) ; General purpose PVC; C_a=0.94

Table A5(1)

Correction factors for ambient temperature

Note: This table applies where the associated overcurrent protective device is intended to provide short circuit protection only. Except where the device is a semi-enclosed fuse to BS3036 the table also applies where the device is intended to provide overload protection.

Type of insulation	Operating temperature	Ambient Temperature °C														
		25	30	35	40	45	50	55	60	65	70	75	80	85	90	95
Rubber (flexible cables only)	60°C	1.04	1.0	0.91	0.82	0.71	0.58	0.41	—	—	—	—	—	—	—	—
General purpose PVC	70°C	1.03	1.0	0.94	0.87	0.79	0.71	0.61	0.50	0.35	—	—	—	—	—	—
Paper	80°C	1.02	1.0	0.95	0.89	0.84	0.77	0.71	0.63	0.55	0.45	0.32	—	—	—	—
Rubber	85°C	1.02	1.0	0.95	0.90	0.85	0.80	0.74	0.67	0.60	0.52	0.43	0.30	—	—	—

3- *Installation Reference Method*

Table A6(1) Method 3

TABLE A6(1)

Single-core PVC insulated cables, non-armoured, with or without sheath

(COPPER CONDUCTORS)

BS 6004
BS 6231
BS 6346

CURRENT CARRYING CAPACITY (Amperes):

Ambient temperature: 30°C
Conductor operating temperature: 70°C

Conductor cross-sectional area	Reference Method 4 (enclosed in conduit in thermally insulating wall etc.)		Reference Method 3 (enclosed in conduit on a wall or in trunking etc.)		Reference Method 1 (clipped direct)		Reference Method 11 (on a perforated cable tray horizontal or vertical)		Reference Method 12 (free air)		
									Horizontal flat spaced	Vertical flat spaced	Trefoil
1	2 cables, single-phase a.c. or d.c. 2	3 or 4 cables three-phase a.c. 3	2 cables, single-phase a.c. or d.c. 4	3 or 4 cables three-phase a.c. 5	2 cables, single-phase a.c. or d.c. flat and touching 6	3 or 4 cables three-phase a.c. flat and touching or trefoil 7	2 cables, single-phase a.c. or d.c. flat and touching 8	3 or 4 cables three-phase a.c. flat and touching or trefoil 9	2 cables, single-phase a.c. or d.c. or 3 cables three-phase a.c. 10	2 cables, single-phase a.c. or d.c. or 3 cables three-phase a.c. 11	3 cables trefoil three phase a.c. 12
mm²	A	A	A	A	A	A	A	A	A	A	A
1	11	10.5	13.5	12	15.5	14	—	—	—	—	—
1.5	14.5	13.5	17.5	15.5	20	18	—	—	—	—	—
2.5	20	18	24	21	27	25	—	—	—	—	—
4	26	24	32	28	37	33	—	—	—	—	—
6	34	31	41	36	47	43	—	—	—	—	—
10	46	42	57	50	65	59	—	—	—	—	—
16	61	56	76	68	87	79	—	—	—	—	—
25	80	73	101	89	114	104	131	114	146	130	110
35	99	89	125	110	141	129	162	143	181	162	137
50	119	108	151	134	182	167	196	174	219	197	167
70	151	136	192	171	234	214	251	225	281	254	216
95	182	164	232	207	284	261	304	275	341	311	264
120	210	188	269	239	330	303	352	321	396	362	308
150	240	216	300	262	381	349	406	372	456	419	356
185	273	245	341	296	436	400	463	427	521	480	409
240	321	286	400	346	515	472	546	507	615	569	485
300	367	328	458	394	594	545	629	587	709	659	561
400	—	—	546	467	694	634	754	689	852	795	656
500	—	—	626	533	792	723	868	789	982	920	749
630	—	—	720	611	904	826	1 005	905	1 138	1 070	855
800	—	—	—	—	1 030	943	1 086	1 020	1 265	1 188	971
1000	—	—	—	—	1 154	1 058	1 216	1 149	1 420	1 337	1 079

Examples

Design current i/. $I_b = \frac{P}{U_o}$

$$I_b = \frac{7600}{230} = \underline{33.04A}$$

Rating of the protective device ii/. $I_n = \underline{40A}$ (next size up)

Current carrying capacity of cable iii/. $I_t = \frac{I_n}{\text{Correction factors}}$

$$I_t = \frac{40}{0.94} = \underline{42.55A}$$

Tabulated current carrying capacity of cable (Table A6(1) BS7671)

iv/. $I_t = \underline{57A} = \underline{10mm^2}$

Volt drop $= mV/A/m \times I_b \times$ length

Volt drop $= 4.4 \times 10^{-3} \times 33.04 \times 25 = \underline{3.63V}$

- Example

An immersion heater rated at 240 V, 3 kW is to be installed using twin with protective conductor p.v.c. insulated and sheathed cable. The circuit will be fed from a 15 A miniature circuit breaker type 2, and will be run for much of its 14 m length in a roof space which is thermally insulated with glass fibre. The roof space temperature is expected to rise to 50°C in summer, and where it leaves the consumer unit and passes through a 50 mm insulation-filled cavity, the cable will be bunched with seven others. Calculate the cross-sectional area of the required cable

1) First calculate the design current Ib

Ib=P/V=3000/240=12.5A

In=15 A

2) The ambient temperature correction factor is found from {Table 4.3} to be Ca=0.71.

3) The group correction factor is found from {Table 4.4} as Cg=0.52.

(The circuit in question is bunched with seven others, making eight in all).

4) We must consider the point where the bunched cables pass through the insulated cavity. From {Table 4.5} we have a factor of Ci=0.89.

5) If we consider all correction factors, It=In/(Cg*Ca*Ci)=15/(0.89*0.71*0.52)=46.65A

6) But we usually consider only minimum of Ca and Cg*Ci. **The combined value of Cg*Ci is (0.463), which is lower than the ambient temperature correction factor of 0.71, and will thus be the figure to be applied. Hence the required current rating is calculated:** It=In/(Cg*Ci)=40/(Cg*Ci)=15/(**0.52 x 0.89**)=32.4 A

6) From table A 4.7, **we can choose 6 mm² p.v.c. twin with protective conductor has a current rating of 32 A.**

7) The table concerned here is {4.7}, which shows a figure of 7.3 mV/A/m for 16 mm² twin with protective conductor pvc insulated and sheathed cable. The actual circuit current is 12.5 A, and the length of run is 14 m.

Voltage drop=7.3*12.5*14/1000=1.28 V

8) Maximum permissible voltage drop=4% of 240 V=9.6 V.

9)Maximum allowable cable length:

If a 14 m run gives a volt drop of 1.28 V, the length of run for a 9.6 V drop will be= **9.6 x 14m/1.28=** 105 m

Table 4.7 - Current ratings and volt drops for sheathed multi-core p.v.c.-insulated cables

Cross sectional area	conduit in thermal insulation	conduit in thermal insulation	In conduit on wall	In conduit on wall	Clipped direct	Clipped direct	Volt drop	Volt drop
(mm^2)	(A)	(A)	(A)	(A)	(A)	(A)	(mV/A/m)	(mV/A/m)
-	2 core	3 or 4 core	2 core	3 or 4 core	2 core	3 or 4 core	2 core	3 or 4 core
1.0	11.0	10.0	13.0	11.5	15.0	13.5	44.0	38.0
1.5	14.0	13.0	16.5	15.0	19.5	17.5	29.0	25.0
2.5	18.5	17.5	23.0	20.0	27.0	24.0	18.0	15.0
4.0	25.0	23.0	30.0	27.0	36.0	32.0	11.0	9.5
6.0	32.0	29.0	38.0	34.0	46.0	41.0	7.3	6.4
10.0	43.0	39.0	52.0	46.0	63.0	57.0	4.4	3.8
16.0	57.0	52.0	69.0	62.0	85.0	76.0	2.8	2.4

- Example:

Assume that the immersion heater indicated in the two previous examples is to be installed, but this time with the protection of a 15 A rewirable (semi-enclosed) fuse. Calculate the correct cable size for the cable if it does not run in thermal insulation.

Ca= **0.71**

Cg=0.52

Cf=0.725

It=In/(Ca*Cg*Cf)=15/(0.71*0.52*0.725)=56 A

The correct size for 56 A will be 16 mm².

Mineral insulated cables

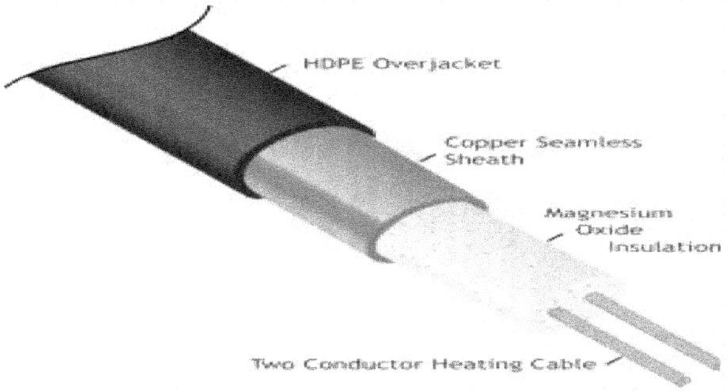

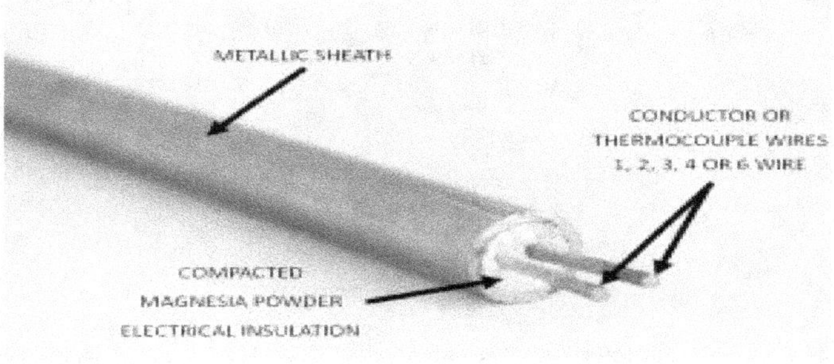

Example:

A 415 V 50 Hz three-phase motor with an output of 7.5 kW, power factor 0.8 and efficiency 85% is the be wired using 500 V light duty three-core mineral insulated p.v.c. sheathed cable. The length of run from the HBC protecting fuses is 20 m, and for about half this run the cable is clipped to wall surfaces. For the remainder it shares a cable tray, touching two similar cables across the top of a boiler room where the ambient temperature is 50°C. Calculate the rating and size of the correct cable.

1)The first step is to calculate the line current of the motor.

Pout/Pin=eff

Pin=Pout/eff= 7.5*1000/0.85=8.82 kW

2) The rated current Ib

Ib=Pin/(1.73*Vl*pf)

$$I_b = \frac{P_{in}}{1.73*V_l*Pf} = \frac{8.82*1000}{1.73*415*0.8} = 15.3$$

1)Now we select suitable fuse. We must now select a suitable fuse. {Fig 3.15} for BS 88 fuses shows the 16 A size to be the most suitable.

2)Part of the run is subject to an ambient temperature of 50°C, where the cable is also part of a group of three, so the appropriate correction factors must be applied from {Tables 4.3 and 4.4}. Note that the grouping factor of 0.70 has been selected because where the cable is grouped it is clipped to a metallic cable tray, and not to a non-metallic surface.

$I_t = I_n/(C_g*C_a) = 16/(0.7*0.67) = 34.2$ A

3) Next the cable must be chosen from {Table 4.8}. Since part of the cable is on the tray (method 3) the correct size for 34.2 A will be 4.0 mm^2.

4) This time we have a mineral insulated p.v.c. sheathed cable, so volt drop figures will come from {Table 4.9}. This shows 9.1 mV/A/m for the 4 mm^2 cable selected, which must be used with the circuit current of 15.3 A and the length of run which is 20 m.

Voltage drop=9.1*15.3*20/1000=2.78 V

5) Maximum permissible voltage drop=4% of 415 V=16.6 V.

6)Maximum allowable cable length:

Maximum length of run for this circuit with the same cable size and type will be:16.6*20/2.78=119m

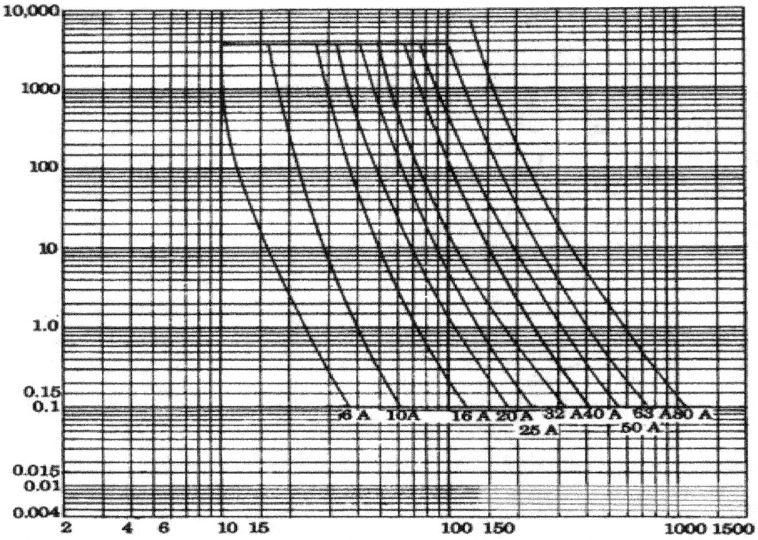

Table 4.8 - Current ratings of mineral insulated cables clipped direct

Cross-sectional area	Volt	p.v.c. sheath 2 x single or twin	p.v.c. sheath 3 core	p.v.c. sheath 3 x single or twin	Bare sheath 2 x single	Bare sheath 3 x single
(mm²)		(A)	(A)	(A)	(A)	(A)
1.0	500v	18.5	16.5	16.5	22.0	21.0
1.5	500v	24.0	21.0	21.0	28.0	27.0
2.5	500v	31.0	28.0	28.0	38.0	36.0
4.0	500v	42.0	37.0	37.0	51.0	47.0

Table 4.9 - Volt drops for mineral insulated cables

Cross-sectional area	Single-phase p.v.c. sheath	Single-phase bare	Three-phase p.v.c. sheath	Three-phase bare
(mm²)	(mV/A/m)	(mV/A/m)	(mV/A/m)	(mV/A/m)
1.0	42.0	47.0	36.0	40.0
1.5	28.0	31.0	24.0	27.0
2.5	17.0	19.0	14.0	16.0
4.0	10.0	12.0	9.1	10.0
6.0	7.0	7.8	6.0	6.8
10.0	4.2	4.7	3.6	4.1
16.0	2.6	3.0	2.3	2.6

Application of rating factors

Some or all of the onerous conditions just outlined may affect a cable along its whole length or parts of it, but not all may affect it at the same time. So, consider the following:

1. If the cable in Fig. 18.1 ran for the whole of its length, grouped with others of the same size in a high ambient temperature, and was totally surrounded with thermal insulation, it would seem logical to apply all the CFs, as they all affect the whole cable run. Certainly the factors for the BS 3036 fuse, grouping and thermal insulation should be used. However, it is doubtful if the ambient temperature will have any effect on the cable, as the thermal insulation, if it is efficient, will prevent heat from reaching the cable. Hence apply C_g, C_f and C_i.

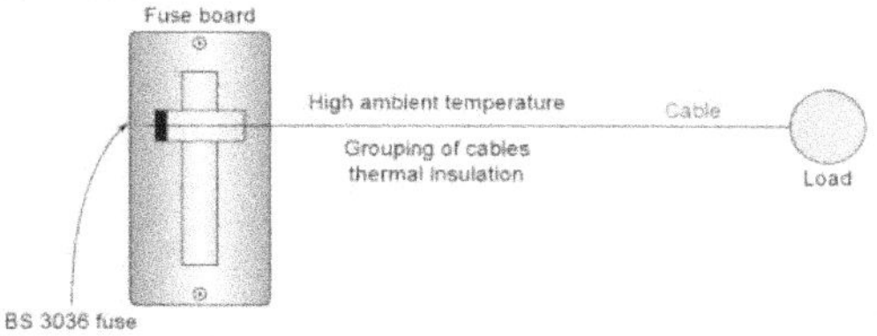

Figure 18.1

2. In Fig. 18.2a the cable first runs grouped, then leaves the group and runs in high ambient temperature, and finally is enclosed in thermal insulation. We therefore have three different conditions, each affecting the cable in different areas. The BS 3036 fuse affects the whole cable run and therefore C_f must be used. However there is no need to apply all of the remaining factors as the worse one will automatically compensate for the others. The relevant factors are shown in Fig. 18.2b: apply only $C_f = 0.725$ and $C_i = 0.5$. If protection was **not** by BS 3036 fuse, then apply only $C_i = 0.5$.

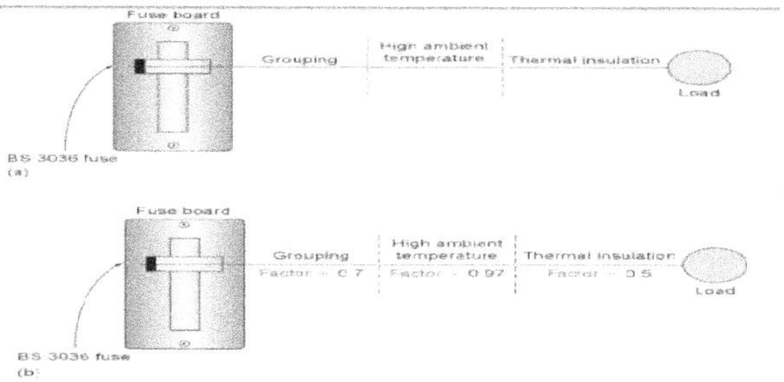

3. In Figs 18.3 and 18.4, a combination of cases 1 and 2 is considered. The effect of grouping and ambient temperature is 0.7 × 0.97 = 0.69. The factor for thermal insulation is still worse than this combination, and therefore C_i is the only one to be used.

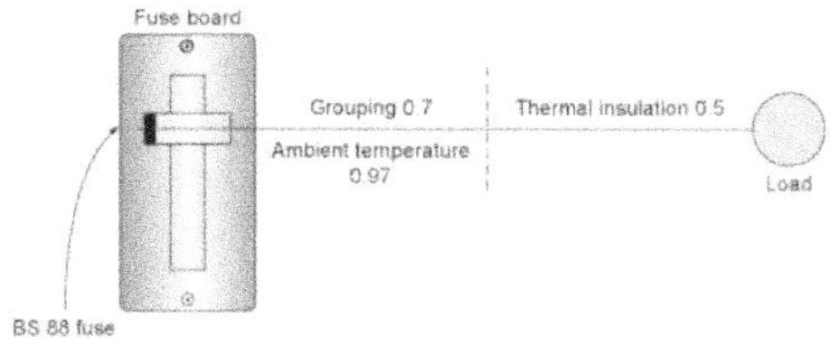

Fi

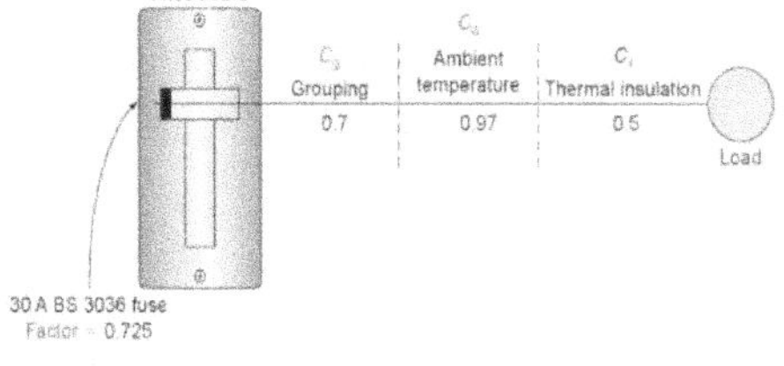

Figure 18.4

Current-carrying capacity

The required formula for tabulated current-carrying capacity I_t is

$$I_t \geq \frac{I_n}{\text{relevant CFs}}$$

In Fig. 18.4, the current-carrying capacity is given by

$$I_t \geq \frac{I_n}{C_f C_i} \geq \frac{30}{0.725 \times 0.5} \geq 82.75\ \text{A}$$

or, without the BS 3036 fuse,

$$I_t \geq \frac{30}{0.5} \geq 60\ \text{A}$$

In Fig. 18.5, $C_a C_g = 0.97 \times 0.5 = 0.485$, which is worse than C_i (0.5). Hence

$$I_t \geq \frac{I_n}{C_f C_a C_g} \geq \frac{30}{0.725 \times 0.485} \geq 85.3\ \text{A}$$

or, without the BS 3036 fuse,

$$I_t \geq \frac{30}{0.485} \geq 61.85\ \text{A}$$

Choice of cable size

Having established the tabulated current-carrying capacity I_t of the cable to be used, it is now essential to choose a cable to suit that value. The tables in appendix 4 of the IET Regulations list all the cable sizes, current-carrying capacities and voltage drops of the various types of cable. For example, in the case of single core 70°C thermoplastic insulated cables, which are single phase, in conduit and have a current-carrying capacity of 45 A, the installation is by reference method B (Table 4A2), the cable table is 4D1A and the column is 4. Hence the cable size is 10.0 mm² (column 1).

Voltage drop

The resistance of a conductor increases as the length increases and/or the cross-sectional area (c.s.a.) decreases. Associated with an increased resistance is a drop in voltage, which means that a load at the end

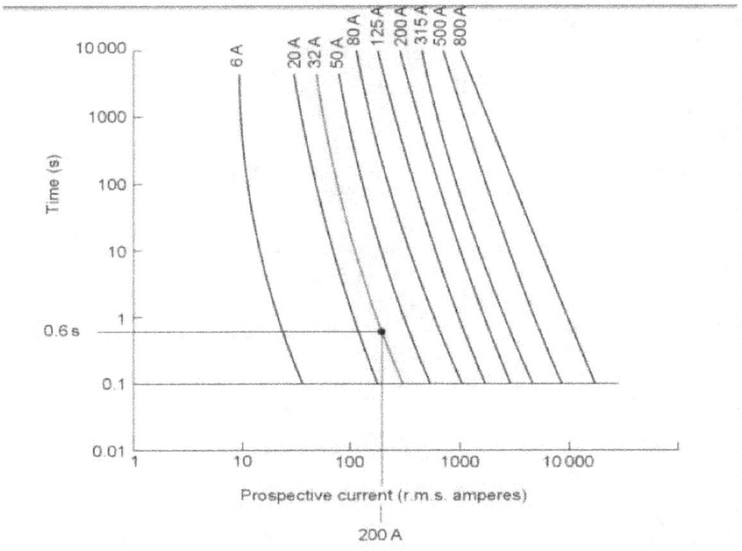

Figure 18.8 Time/current characteristics.

• http://www.melissaelectrical.co.uk/services/cable-calculator/

LO4

Apply short-circuit and over-load protection principles for electrical installations.

LO4-1

Over Current Protection

- **An over current is the increase of electrical current in the system above the level for which it is designed.**
- **Whenever the current passes through the cable, heat is generated *but* it should be kept within the limits.**
- **Over current is one of the major safety hazards**
- **Due to over current, there are chances of Electric shocks**
- **Over currents lead to risk of fire**
- **In UK every year about 50,000 fires are due to electrical faults**

3

Over Current Protection

Types of over-current

1. Overload

➢ An overload occurs when a current flows that is somewhat too high (usually 50% to 100% too high) for the system. Overloads don't normally cause immediate, catastrophic damage. Instead, the likelihood of damage increases gradually as the duration of the overload increases. If the fault is not resolved, cables will overheat and melt, exposing bare conductors. The heat generated may be sufficient to cause a fire.

➢ In a domestic setting, overloads usually result from using too many appliances at the same time, or plugging a heavy-duty appliance into a supply that isn't strong enough for it. An example of the latter is connecting an electric shower to a standard 13-amp plug, and plugging it into a socket.

4

Types of over-current

2. Short-circuit

➢ A short-circuit is a connection between live and neutral, or between live and earth, that bypasses an appliance. The connection will probably have a low resistance, and the current that can flow may be hundreds or thousands of times too high for the system. This current is usually called the *fault current or short-circuit current.*

➢ The ability to handle short-circuits is not just important to protect cables, it is part of the protection against electric shock. If a live conductor in, say, an electric kettle becomes loose and touches the metal case, we hope that a large fault current will flow. This current will flow from the live, through the case, and back to earth via the earth wire. The fault current will blow the fuse or trip the MCB, thus rendering the circuit dead.

➢ In practice, in domestic installations overload protection and short-circuit protection are both provided by the same device: either a fuse of an MCB. Additional shock protection may be provided by an RCD. Whether a fuse or an MCB is used, when the current exceeds a certain limit for a certain time, the fuse will `blow' (break) or the MCB will `trip'. In both cases this will open the circuit and prevent the flow of further current.

Fuses

1- Introduction

The fuse is the oldest, simplest and least expensive type of electrical protection device. Its operation is simple: excessive current creates thermal energy, which causes a fuse element to melt, interrupting the path of electrical current flowing through it see figures 1 & 2.

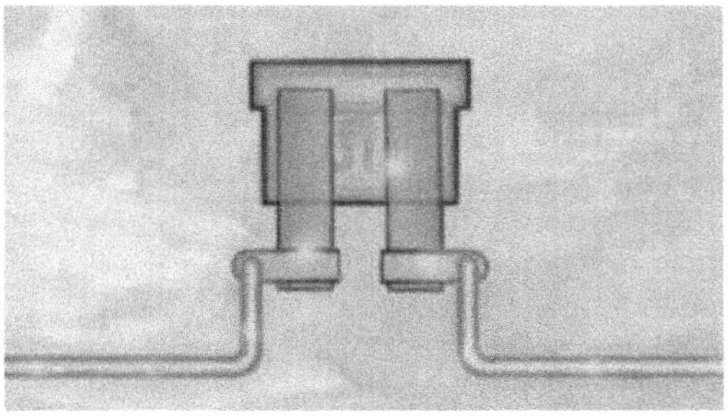

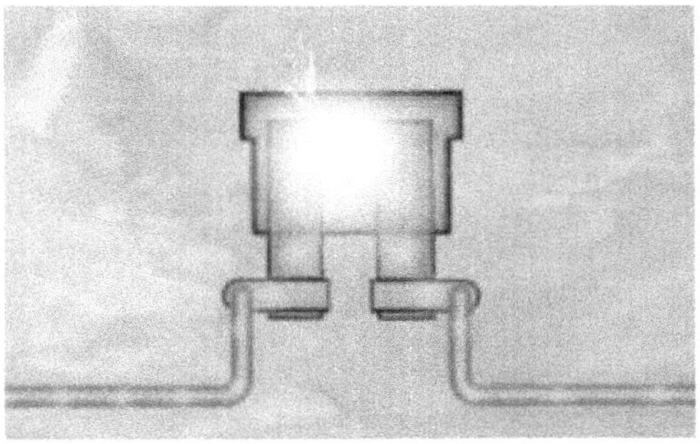

3- Fuse characteristic

The fuse has an inverse characteristic. When the fault current is high, the response time is low. At the fault current If, the fuse starts melting, after a certain time it will get cut. The fault is then cleared.

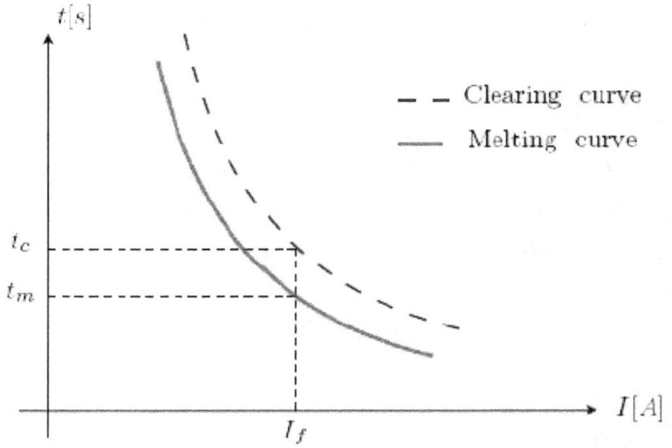

Figure 4.3: Clearing - Melting Curves

4- HRC Fuses

The HRC fuse consists of a ceramic body usually of steatite, pure silver element, clean silica quartz, asbestos washers, porcelain plugs, brass end caps and copper tags see Figure (4.6)

a- High rupturing capacity (HRC) fuses are used in low voltage situations where fault currents are high. The insulating barrel of the fuse cartridge contains an arc extinguishing compound [7].

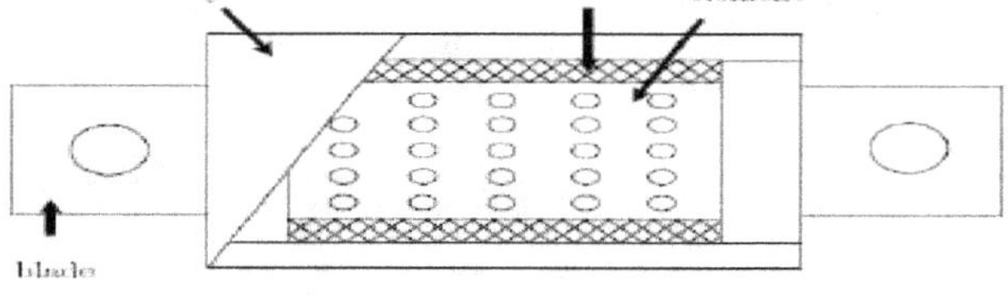

Figure 4.6: Fuse Structure

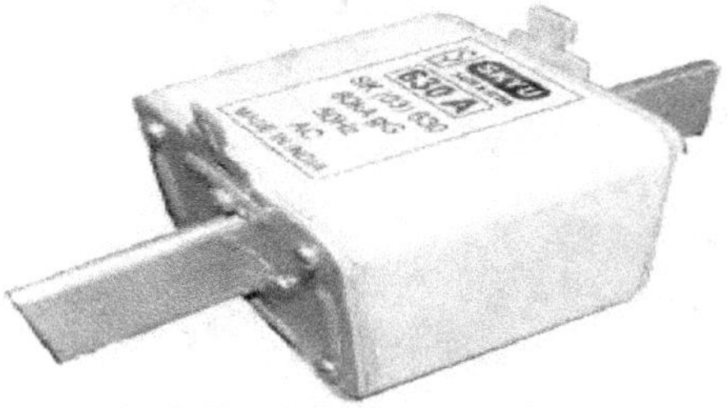

Typical rupturing capacities:

- Re-wirable up to 4 kA
- HRC types up to 100 + kA

The standard current ratings for low voltage cartridge fuses in amperes are:

2 - 4 - 6 - 8 - 10 - 12 - 16 - 20 – 25 - 32 - 40 - 50 - 63 - 80 - 100 – 125 - 160 – 200 - 250 - 315 - 400 - 500 - 630 - 800 - 10000 - 10250

b- High Voltage Drop out Fuses are commonly used to protect pole mounted distributed transformers. A barrel constructed of an insulating

Fig 1: Medium voltage fuse for transformer protection, 24 kV, 200 A

Fig 2: High voltage fuse, 3 phases 115kV

- A cartridge type, used on overhead "high" or" medium" voltage pole lines (6.6 - 22 kV)
- Fuse element in cartridge, which explodes out and can cause bush-fires

Fig.3 Cartridge fuse 10A

c- HRC Fuse element during high current interruption

- During the pre-arc time the fuse element remains a simple resistance that does not change the circuit characteristics.

- An arc is ignited at each neck (4 arcs in series in this example) see Figure (4.7).
- Melting: The grains of sand in contact with the arc melt and form a glassy body known as "fulgurite".

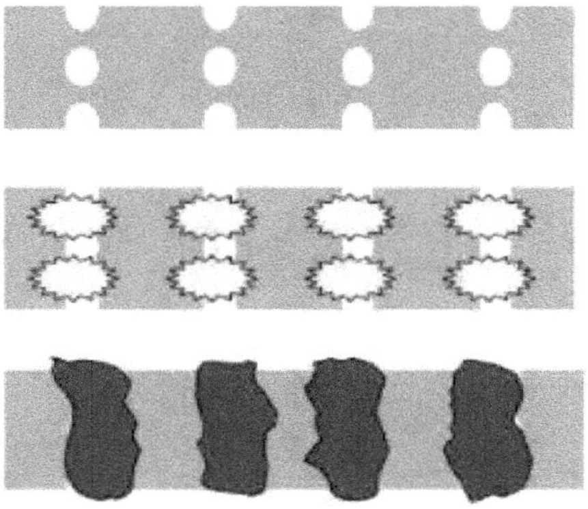

5-. Fuse Rating

It is that value of current which when flows through the element does not melt it.

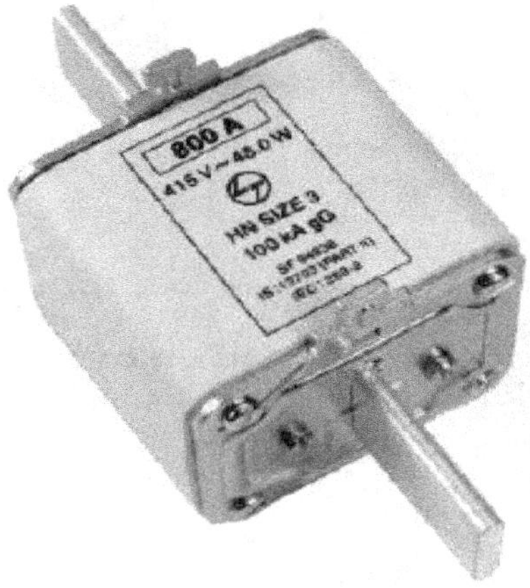

- Nominal voltages, Vn. the voltage the fuse can operate at. The max voltage is usually Vn +10 %. Care! - max voltage must be matched to max circuit voltage

- Current, In. This will dictate the current-melting time characteristic

- Rated Interruption current. The maximum S/C current the device can safely interrupt, at the rated nominal voltage

- AC or DC

- Fusing factor. The ratio of the current which will just cause the fuse to operate to nominal current. Note - definitions vary from country to country.

6- IEC Fuse Classifications

We can see the comparison of the time current curve of 4 different IEC1 fuses rated 100A in Figure (4.8)
Note the use of log-log scales, to accommodate the very wide range in both currents and operating times involved.

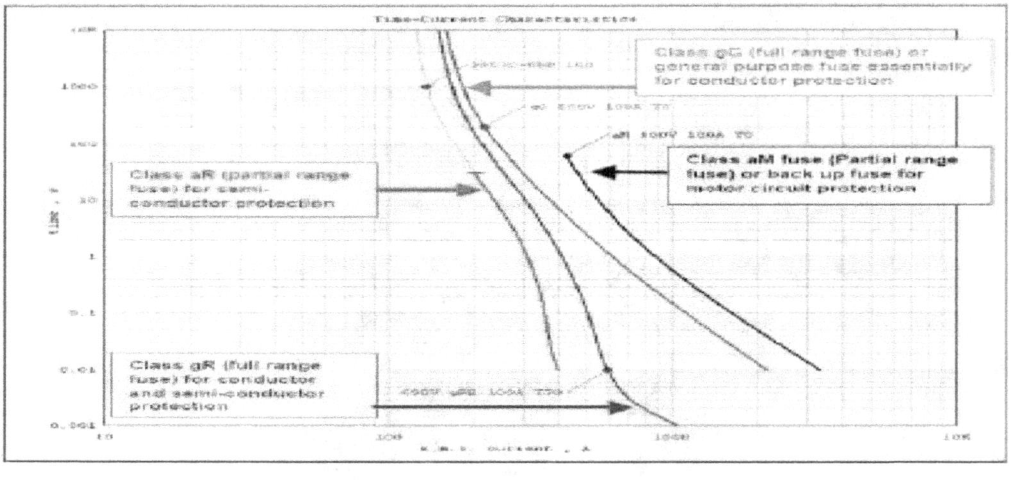

Figure 4.8: Time Current Curve of 4 different IEC Fuses Rated 100A

7- Fuse Types

A fuse type is indicated by two letters. The first letter indicates the main operating mode and the second letter indicates the object to be protected as shown in Table (4.1).

• The first letter indicates the main operating mode:

a: associated fuse. It must be associated to another protective device as it cannot interrupt faults below a specified level. Short circuit protection only.

g: general purpose fuse. It will interrupt all faults between the lowest fusing current (even if it takes 1 hour to melt the fuse elements) and the breaking capacity. Overload and short circuit protection.

• The second letter indicates the object to be protected:

G: cable and conductor protection, general
M: motor circuit protection
R: semi conductor protection
S: semi conductor protection
Tr: transformer protection
Note that even for the same fuse rating, operating times for high currents can vary considerably depending on type i.e. whether 'fast' (e.g. aR, gR) or 'slow' types (gG, aM). This has grading implications between upstream and downstream fuses,especially if the upstream device is a 'fast' fuse.

Fuse Type	Typical Industrial Applications	Operating Range
aM	Motor circuits protection against short circuit only	Partial Range
aR	IEC 269-4 fuse for semi conductor protection	
gG	General purpose fuse essentially for conductor protection	Full Range
gM	Motor protection	
gN	North American fast acting fuse for general purpose applications, mainly for conductor protection. As per UL 248 class J and class L fuses.	
gD	North American general purpose time-delay fuse for motor circuit protection and conductor protection (for example: fuse class AJT, RK5 and A4BQ)	
gTR	Transformer protection	
gR, gS	IEC 269-4 semi conductor and conductor protection	
gL, gF, gI	Former type of fuses for conductor protection replaced today by the gG fuses	

Table 4.1: Fuse Types

Miniature Circuit Breaker

1- MCB Operation

The operation of the MCB in order to ensure protection against overload currents and short circuit currents is summarized in figure 2 below.

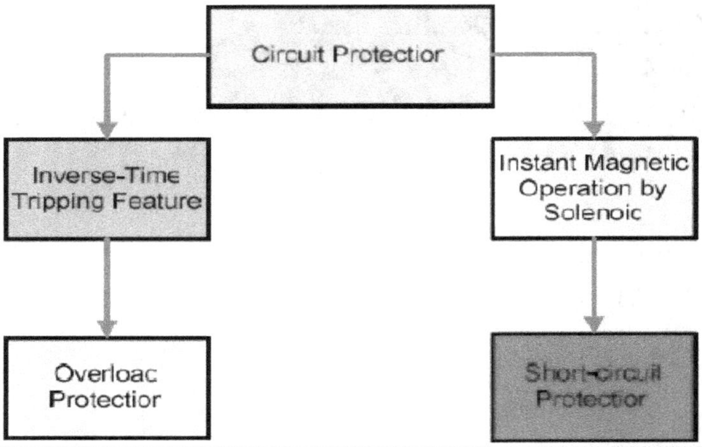

1- Working Principle Miniature Circuit Breaker

There are two arrangement of **operation of miniature circuit breaker**. One due to thermal effect of over current and other due to electromagnetic effect of over current.

a- Thermal effect of over

The thermal **operation of miniature circuit breaker** is achieved with a bimetallic strip whenever continuous over current flows through MCB, the bimetallic strip is heated and deflects by bending. This deflection of bimetallic strip releases mechanical latch. As this mechanical latch is attached with operating mechanism, it causes to open the miniature circuit breaker contacts.

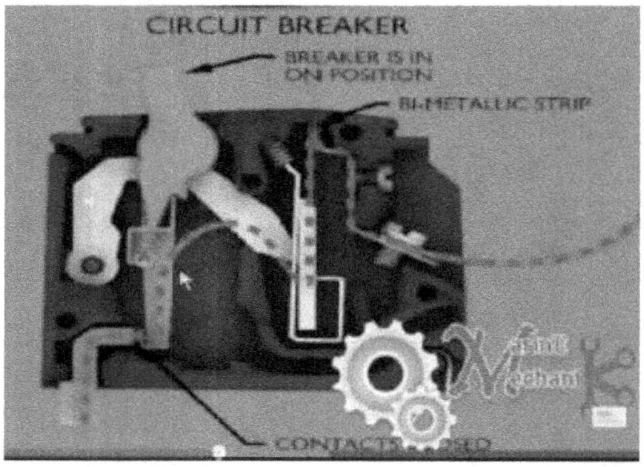

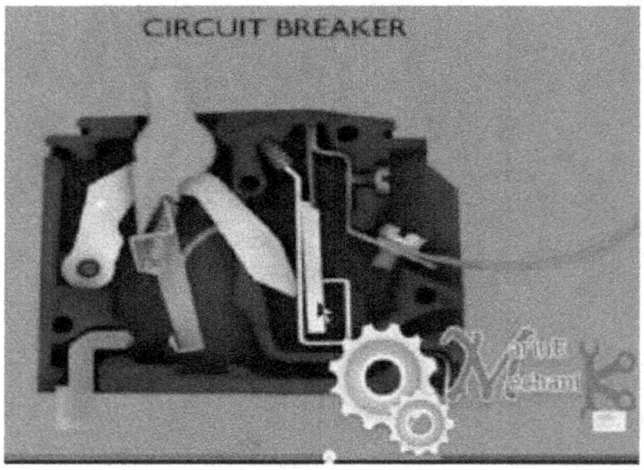

b- electromagnetic effect of over current

But during short circuit condition, sudden rising of current, causes electromechanical displacement of plunger associated with tripping coil or solenoid of MCB. The plunger strikes the trip lever causing immediate release of latch mechanism consequently open the circuit breaker contacts. This was a simple explanation of **miniature circuit breaker working principle**.

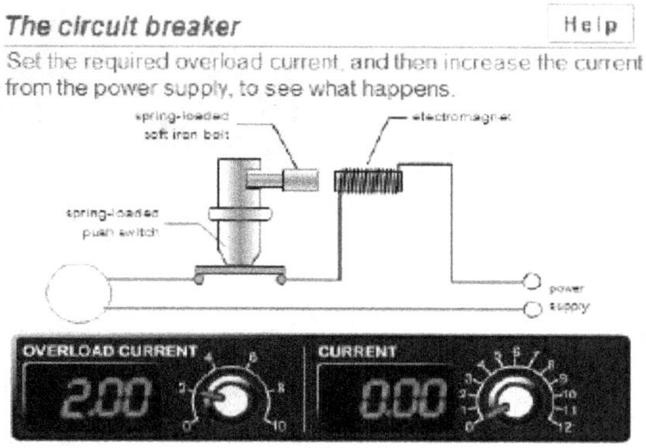

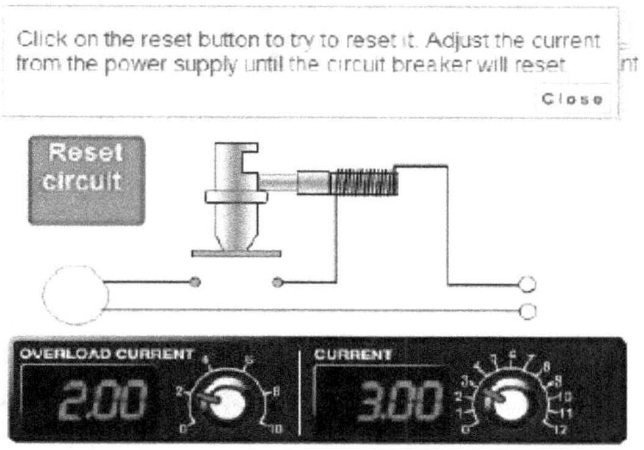

2- MCB Characteristics

The most essential feature of the MCB is the inverse-time tripping characteristic. This feature indicates the time required to trip the breaker in order to clear the circuit of any given level of overcurrent load. A typical inverse time tripping characteristic is depicted in figure 1 below.

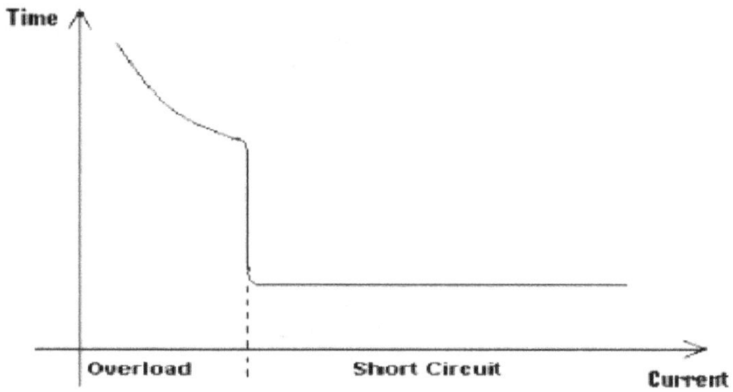

Figure 1: Inverse-time tripping feature of the MCB

4- Miniature Circuit Breaker Construction

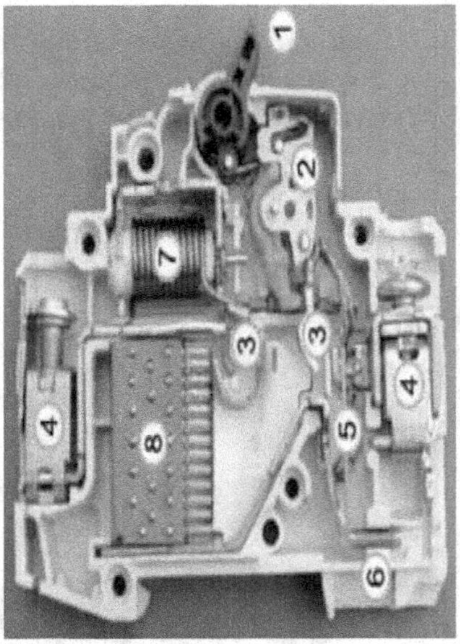

Miniature circuit breaker construction is very simple, robust and maintenance free. Generally a MCB is not repaired or maintained, it just replaced by new one when required. A miniature circuit breaker has normally three main constructional parts. These are:

Frame of Miniature Circuit Breaker

The frame of miniature circuit breaker is a molded case. This is a rigid, strong, insulated housing in which the other components are mounted.

Operating Mechanism of Miniature Circuit Breaker

The operating mechanism of miniature circuit breaker provides the means of manual opening and closing operation of miniature circuit breaker. It has three-positions "ON," "OFF," and "TRIPPED". The external switching latch can be in the "TRIPPED" position, if the MCB is tripped due to over-current. When manually switch off the MCB, the switching latch will be in "OFF" position. In close condition of MCB, the switch is positioned at "ON". By observing the positions of the switching latch one can determine the condition of MCB whether it is closed, tripped or manually switched off.

5- Trip Unit of Miniature Circuit Breaker

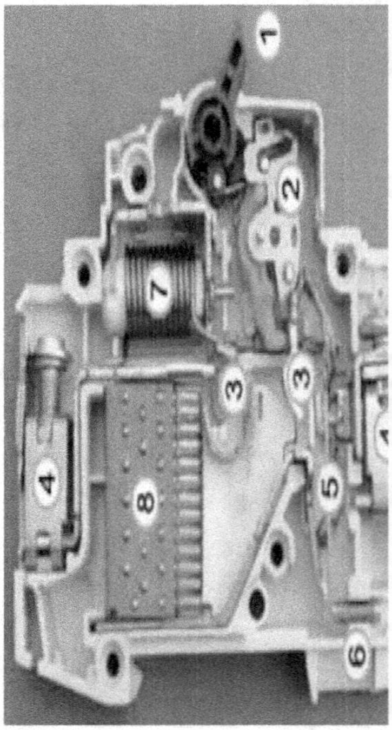

The trip unit is the main part, responsible for proper **working of miniature circuit breaker**. Two main types of trip mechanism are provided in MCB. A bimetal provides protection against over load current and an electromagnet provides protection against short-circuit current.

The DIN rail-mounted thermal-magnetic miniature circuit breaker is the most common style in modern domestic consumer units and commercial electrical distribution boards throughout Europe. The design includes the following components:

1. Actuator lever - used to manually trip and reset the circuit breaker. Also indicates the status of the circuit breaker (On or Off/tripped). Most breakers are designed so they can still trip even if the lever is held or locked in the "on" position. This is sometimes referred to as "free trip" or "positive trip" operation.
2. Actuator mechanism - forces the contacts together or apart.
3. Contacts - Allow current when touching and break the current when moved apart.
4. Terminals
5. Bimetallic strip - separates contacts in response to smaller, longer-term overcurrents
6. Calibration screw - allows the manufacturer to precisely adjust the trip current of the device after assembly.
7. Solenoid - separates contacts rapidly in response to high overcurrents
8. Arc divider/extinguisher

6- **Trip Unit mechanisms of Miniature Circuit Breaker**

a- Overload situation: (bi-metallic strip)

If circuit is overloaded for long time, the bi - metallic strip becomes over heated and deformed. This deformation of bi metallic strip causes, displacement of latch point. The moving contact of the MCB is so arranged by means of spring pressure, with this latch point, that a little displacement of latch causes, release of spring and makes the moving contact to move for opening the MCB.

b- Short circuit situation (trip coil)

The current coil or trip coil is placed such a manner, that during short circuit fault the mmf of that coil causes its plunger to hit the same latch point and make the latch to be displaced. Hence the MCB will open in same manner.

c- Manual operation (hand operated lever)

Again when operating lever of the miniature circuit breaker is operated by hand, that means when we make the MCB at off position manually, the same latch point is displaced as a result moving contact separated from fixed contact in same manner.

So, whatever may be the operating mechanism, that means, may be due to deformation of bi - metallic strip, due to increased mmf of trip coil or may due to manual operation, actually the same latch point is displaced and same deformed spring is released, which ultimately responsible for movement of the moving contact. When the the moving contact separated

Over Current Protection Devices

BASIS	FUSE	CIRCUIT BREAKER
Working Principle	Fuse works on the electrical and thermal properties of the conducting materials.	Circuit breaker works on the Electromagnetism and switching principle.
Reusability	Fuses can be used only once.	Circuit breakers can be used a number of times.
Status indication	It does not give any indication.	It gives an indication of the status
Auxiliary contact	No auxiliary contact is required.	They are available with auxiliary contact.
Switching Action	Fuse cannot be used as as an ON/OFF switch.	The Circuit breaker is used as an ON/OFF switches.
Temperature	They are independent of ambient temperature	Circuit breaker Depends on ambient temperature
Characteristic Curve	The Characteristic curve shifts because of the ageing effect.	The characteristic curve does not shift.

Over Current Protection Devices

Protection	The Fuse provides protection against only power overloads	Circuit breaker provides protection against power overloads and short circuits.
Function	It provides both detection and interruption process.	Circuit breaker performs only interruption. Faults are detected by relay system.
Breaking capacity	Breaking capacity of the fuse is low as compared to the circuit breaker.	Breaking capacity is high.
Operating time	Operating time of fuse is very less (0.002 seconds)	Operating time is comparatively more than that of the fuse. (0.02 – 0.05 seconds)
Version	Only single pole version is available.	Single and multiple version are available.
Mode of operation	Completely automatically.	Manually as well as automatically operated.
Cost	Cost of fuse is low.	Cost of circuit breaker is high.

Over Current Protection Devices

- MCBs are available in various types: `1', `2', `3', `B' and `C'. Each has different characteristics, and is appropriate for a particular application. In a domestic system, we will normally use a type `1' or a type `B' device, as these are general-purpose units.

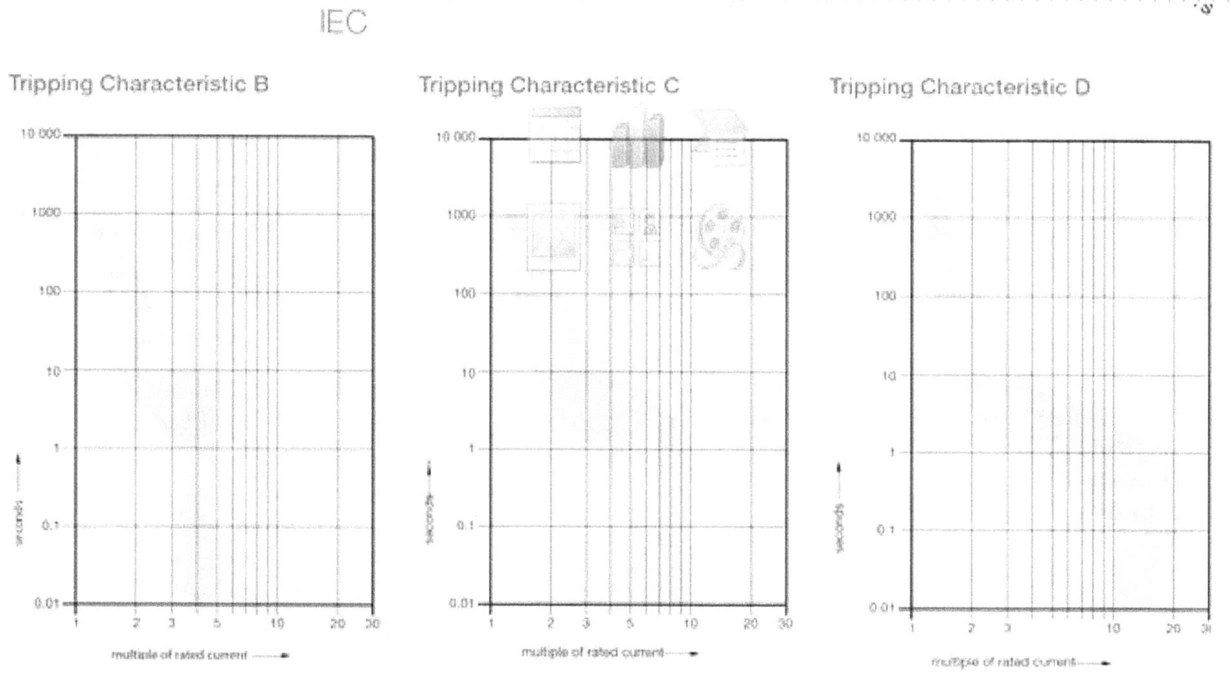

Over Current Protection Devices

'B' curve to BSEN 60898

Magnetic Trip 3 to 5 Times In

For use in domestic applications where the maximum sensitivity is required and there is very little in the way of equipment Which would require a high start up current.

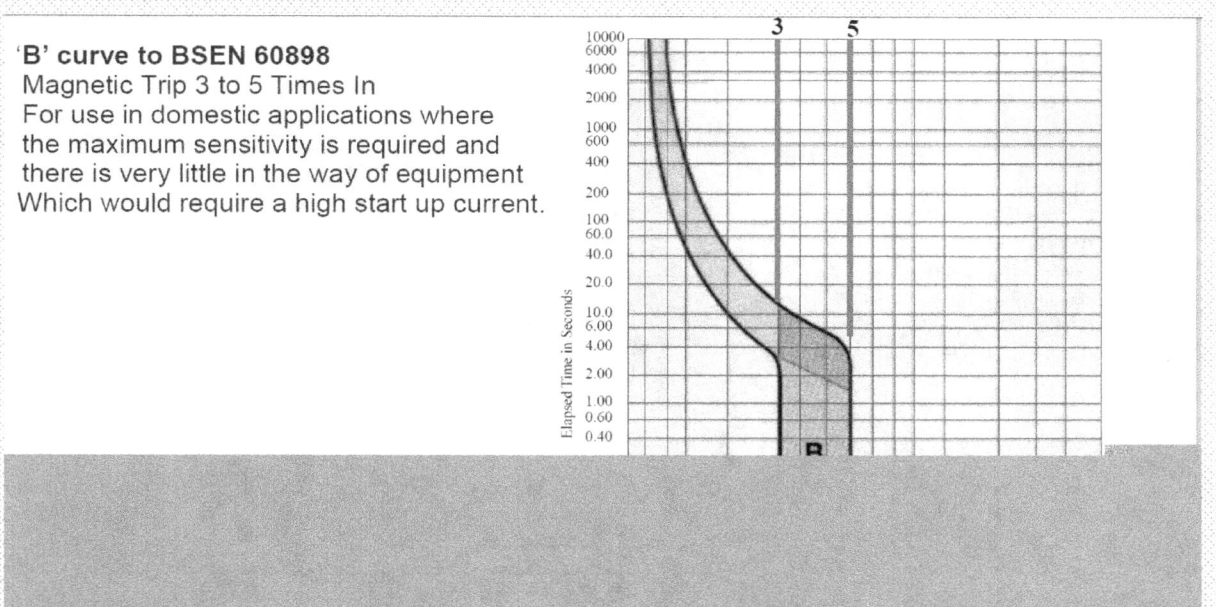

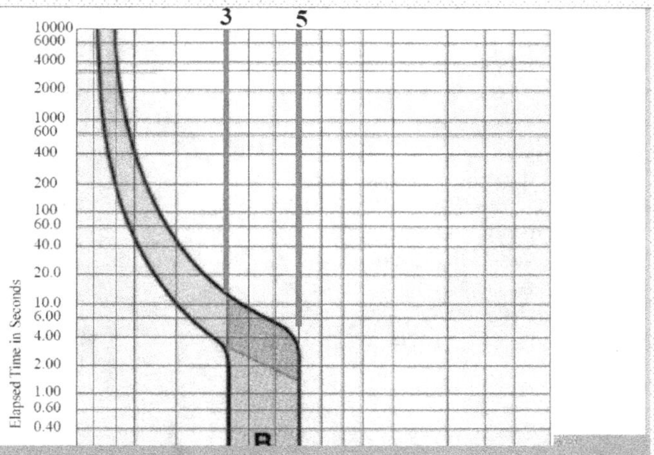

Over Current Protection Devices

'C' Curve to BSEN 60898
Magnetic Trip 5 to 10 Times In

Used in commercial and light industrial applications where close protection is not required and start up currents of devices can run up to 5 times rated current for a short period of time.

If a lot of low voltage lighting is used in a domestic environment then it would prudent to fit C Type mcb due to the inrush current of the transformers thus avoiding nuisance tripping.

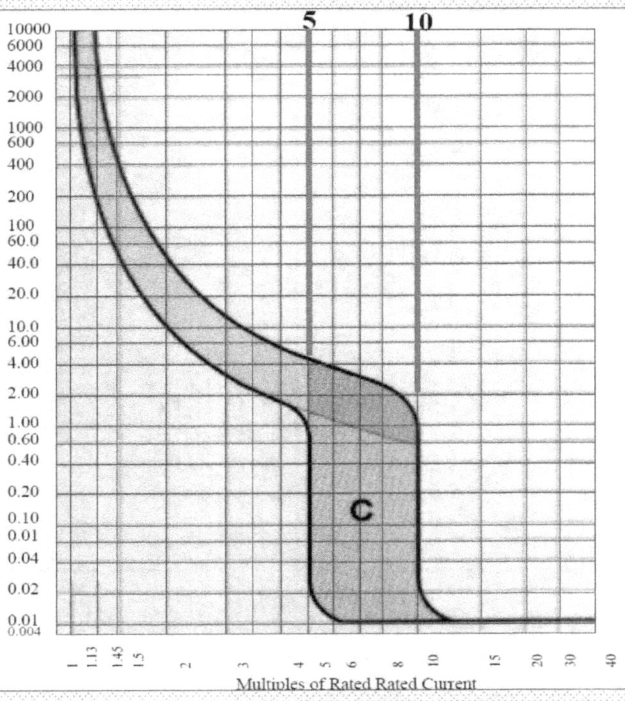

Over Current Protection Devices

'D' Curve to BSEN 60898
Magnetic Trip 10 to 20 Times In

Used in industrial applications where start up current may be 10 times the rated current. For example when using an electric motor with a DOL starter full load current would be at least 7 times the nominal running current.
A large volume of low voltage lighting might also require a D type mcb.

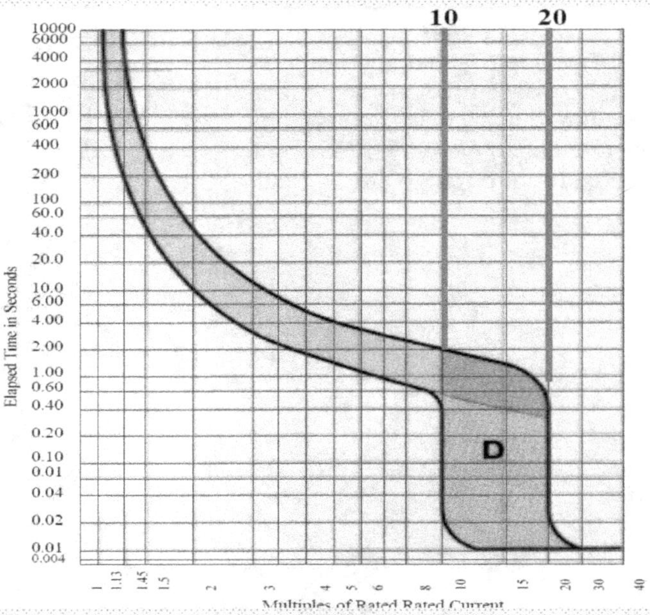

LO4

Apply short-circuit and over-load protection principles for electrical installations.

LO4-2

Grounding or Earthing

- **The process of connecting the metallic frame (i.e. non-current carrying part) of electrical equipment or some electrical part of the system (e.g. neutral point in a star-connected system, one conductor of the secondary of a transformer etc.) to earth (i.e. soil) is called grounding or earthing.**
- **If grounding is done properly, we can effectively prevent accidents and damage to the equipment of the power system and at the same time continuity of supply can be maintained.**

Grounding or Earthing

Grounding or earthing may be classified as:

***1.* Equipment grounding**

2. System grounding

- Equipment grounding deals with earthing the non-current-carrying metal parts of the electrical equipment.
- System grounding means earthing some part of the electrical system *e.g.* earthing of neutral point of star-connected system in generating stations and sub-stations.

1- **Equipment Grounding**

- **The process of connecting non-current-carrying metal parts (i.e. metallic enclosure) of the electrical equipment to earth (i.e. soil) in such a way that in case of insulation failure, the enclosure effectively remains at earth potential is called equipment grounding.**
- **We are frequently in touch with electrical equipment of all kinds, ranging from domestic appliances and hand-held tools to industrial motors.**

Grounding or Earthing

- **Consider a single-phase circuit composed of a 230 V source connected to a motor M as shown in Fig. 1.**
- **Note that neutral is solidly grounded at the service entrance.**

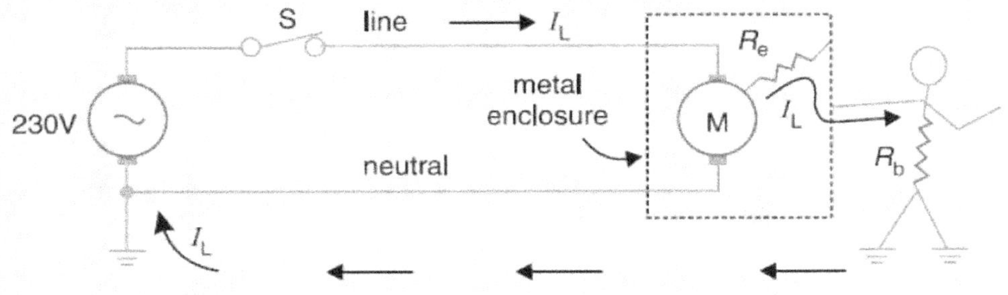

Fig. 1

(i) Ungrounded enclosure.

- **Fig. 1 shows the case of ungrounded metal enclosure. If a person touches the metal enclosure, nothing will happen if the equipment is functioning correctly.**
- **But if the winding insulation becomes faulty, the resistance *Re* between the motor and enclosure drops to a low value (a few hundred ohms or less).**

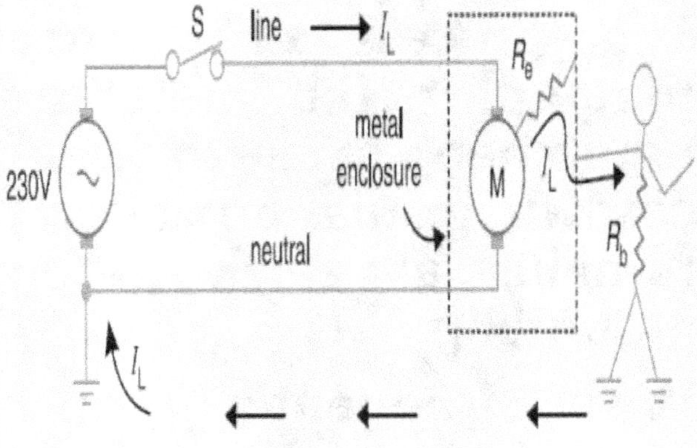

Fig. 1

(ii) Enclosure connected to neutral wire.

- **It may appear that the above problem can be solved by connecting the enclosure to the grounded neutral wire as shown in Fig. 2.**
- **Now the leakage current *IL* flows from the motor, through the enclosure and straight back to the neutral wire.**
- **Therefore, the enclosure remains at earth potential.**
- **The operator would not experience any electric shock .**

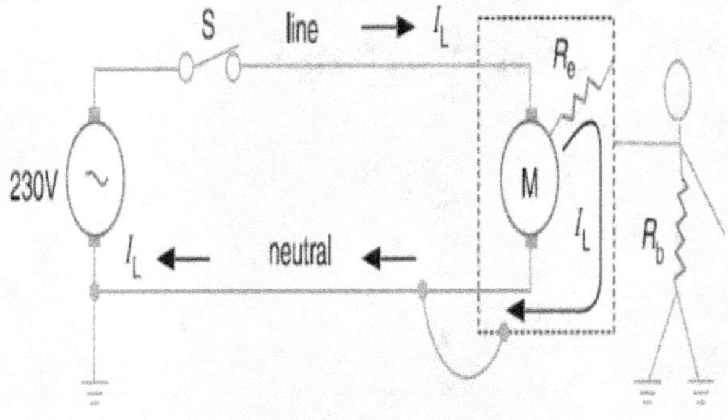

Fig. 2

(ii) Enclosure connected to neutral wire.

- The problem with this method is that the neutral wire may become open either accidentally or due to a faulty installation.
- For example, if the switch is in series with the neutral rather than the live wire (See Fig. 3), the motor can still be turned on and off.
- However, if someone touched the enclosure while the motor is *off,* he would receive a severe electric shock.
- It is because when the motor is off, the potential of the enclosure rises to that of the live conductor.

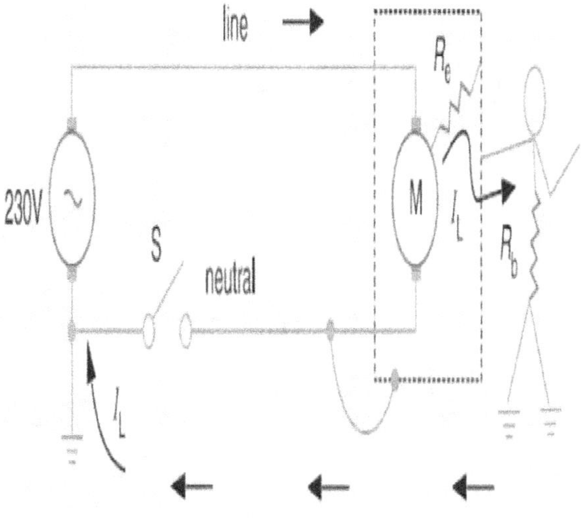

Fig. 3

(iii) Ground wire connected to enclosure

- To get rid of this problem, we install a third wire, called *ground wire,* between the enclosure and the system ground as shown in Fig. 4.
- The ground wire may be bare or insulated.
- If it is insulated, it is colored green.

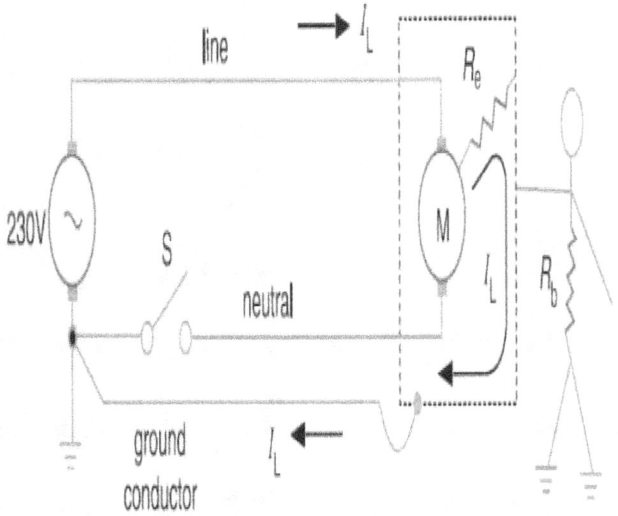

Fig. 4

Electric shock

- It is generally believed that currents below 5 mA are not dangerous.
- Between 10 mA and 20 mA, the current is dangerous because the victim loses muscular control.
- The resistance of the human body, taken between two hands or between one hand and a leg ranges from 500 Ω to 50 kΩ.
- If the resistance of human body is assumed to be 20 kΩ, then momentary contact with a 230 V line can be potentially fatal.
- I_L = 230V/ 20 k Ω = 11.5 mA
- Electrical outlets have three contacts, one for live wire, one for neutral wire and one for ground wire

2- System Grounding

- **The process of connecting some electrical part of the power system (e.g. neutral point of a star connected system, one conductor of the secondary of a transformer etc.) to earth (i.e. soil) is called system grounding.**

2- System Grounding

Example 1

- Fig. 5 (*i*) shows the primary winding of a distribution transformer connected between the line and neutral of a 11 kV line.
- If the secondary conductors are *ungrounded,* it would appear that a person could touch either secondary conductor without harm because there is no ground return.
- However, this is not true.
- Referring to Fig. 5, there is capacitance *C*1 between primary and secondary and capacitance *C*2 between secondary and ground.

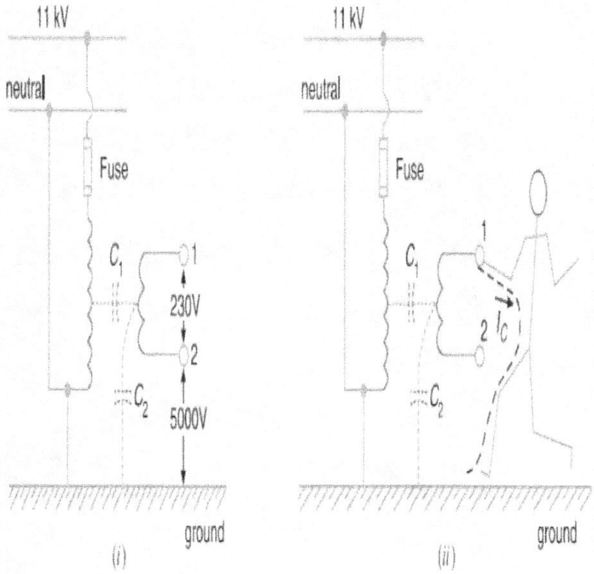

Fig.5

2- System Grounding

- This capacitance coupling can produce a high voltage between the secondary lines and the ground.
- Depending upon the relative magnitudes of *C*1 and *C*2, it may be as high as 20% to 40% of the primary voltage.
- If a person touches either one of the secondary wires, the resulting capacitive current *IC* flowing through the body could be dangerous even in case of small transformers [See Fig. 5(*ii*)].
- For example, if *IC* is only 20 mA, the person may get a fatal electric shock.

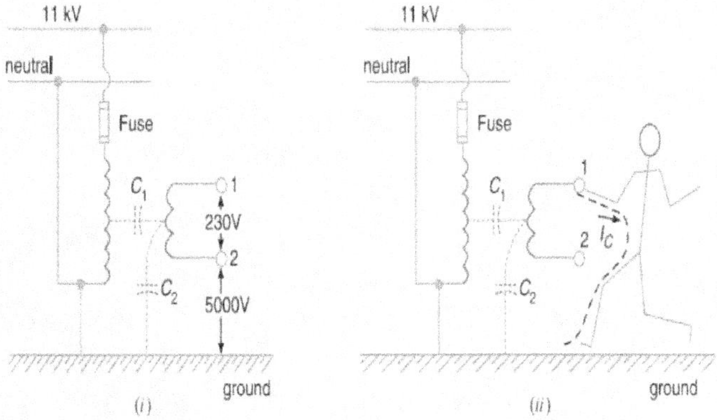

Fig.5

- If one of the secondary conductors is grounded, the capacitive coupling almost reduces to zero and so is the capacitive current *IC*.
- As a result, the person will experience no electric shock.
- This explains the importance of system grounding.

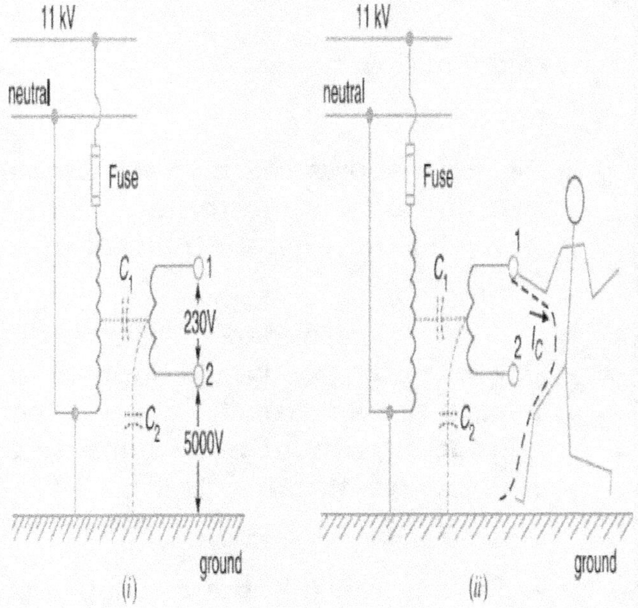

Fig.5

Example 2

- Fig. 6 (*i*) shows the primary winding of a distribution transformer connected between the line and neutral of a 11 kV line.
- The secondary conductors are ungrounded.
- Suppose that the high voltage line (11 kV in this case) touches the 230 V conductor as shown in Fig. 6 (*i*).
- This could be caused by an internal fault in the transformer or by a branch or tree falling across the 11 kV and 230 V lines.
- Under these circumstances, a very high voltage is imposed between the secondary conductors and ground.

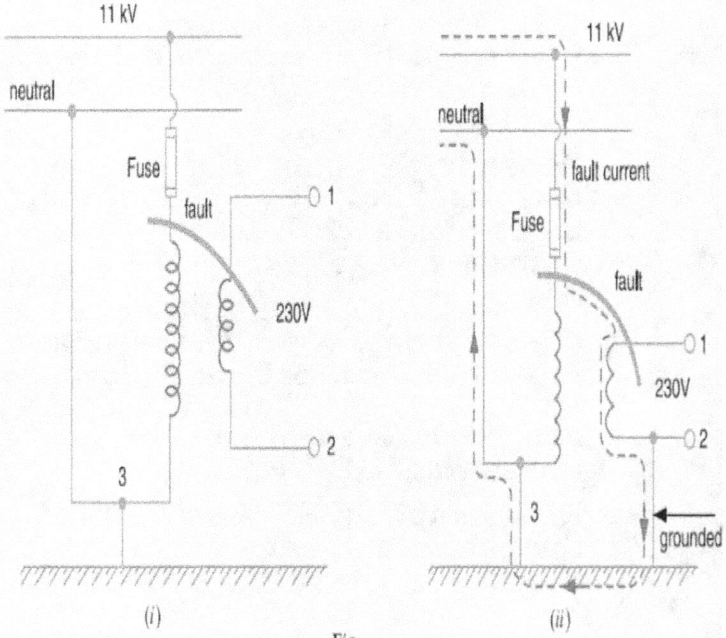

Fig 6

- This would immediately damage the 230 V insulation, causing a massive flashover.
- This flashover could occur anywhere on the secondary network, possibly inside a home or factory.
- Therefore, ungrounded secondary in this case is a potential fire hazard and may produce serious accidents under abnormal conditions.

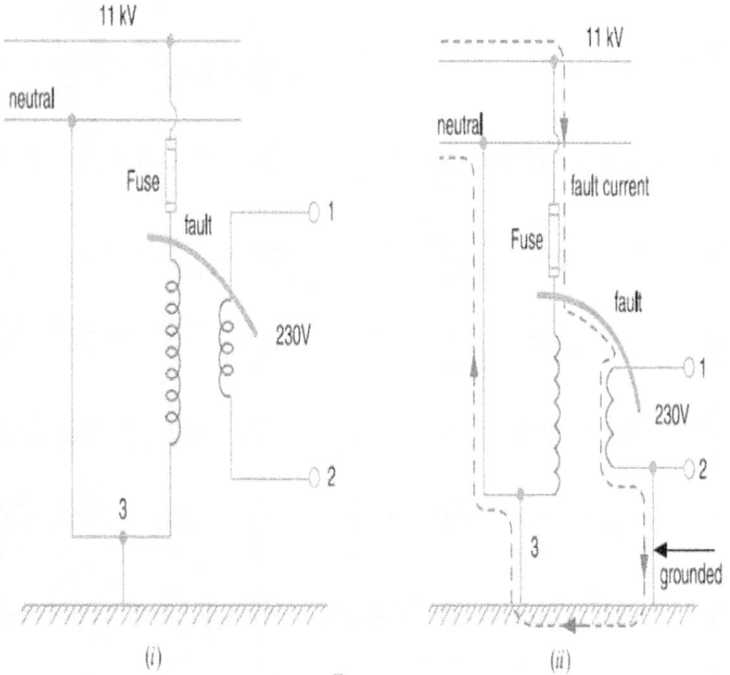

Fig. 6

- If one of the secondary lines is grounded as shown in Fig. 6(*ii*), the accidental contact between a 11 kV conductor and a 230 V conductor produces a dead short.
- The short-circuit current (*i.e.* fault current) follows the dotted path shown in Fig. 6 (*ii*).
- This large current will blow the fuse on the 11 kV side, thus disconnecting the transformer and secondary distribution system from the 11 kV line.
- This explains the importance of system grounding in power system.

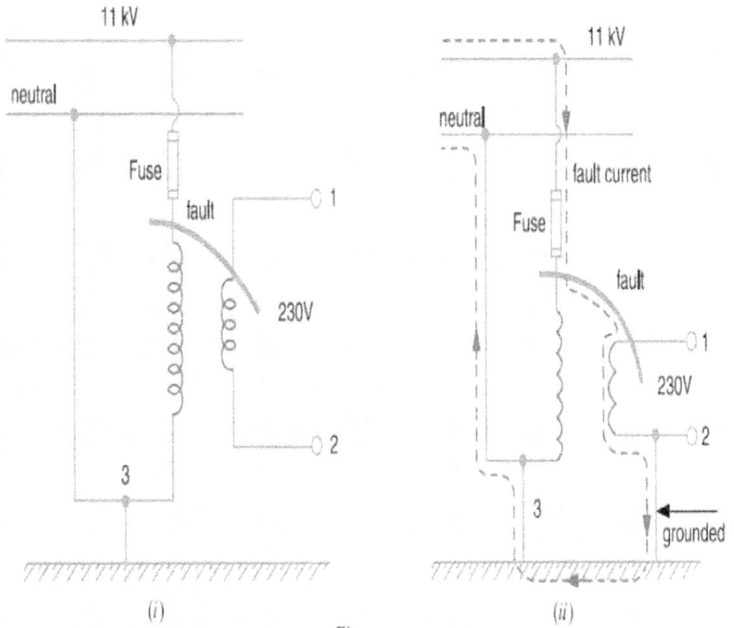

Fig. 6

Earth Leakage Circuit Breaker

- If any current leaks from any electrical installation, there must-be insulation failure in the electrical circuit
- It must be properly detected and prevented otherwise there may be a high chance of electrical shock if-anyone touches the installation.
- An earth leakage circuit breaker does it efficiently.
- It detects the earth leakage current and makes the power supply off by opening the associated circuit breaker.
- There are two types of earth leakage circuit breaker, one is voltage ELCB and other is current ELCB or RCD

1- Voltage Earth Leakage Circuit Breaker

- One terminal of the relay coil is connected to the metal body of the equipment to be protected against earth leakage and other terminal is connected to the earth directly.
- If any insulation failure occurs or live phase wire touches the metal body, of the equipment, there must be a voltage difference appears across the terminal of the relay coil connected to the equipment body and earth.
- This voltage difference produces a current to flow to the relay coil.

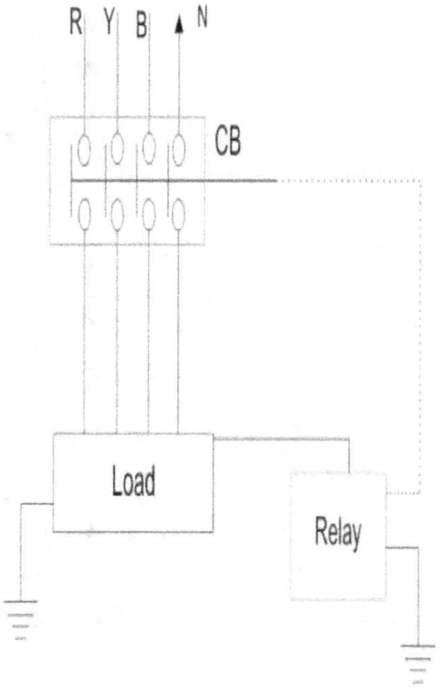

- If the voltage difference crosses, a predetermined limit, the current through the relay becomes sufficient to actuate the relay for tripping the associated circuit breaker to disconnect the power supply to the equipment.
- The typicality of this device is, it can detect and protect only that equipment or installation with which it is attached. It cannot detect any leakage of insulation in other installation of the system.

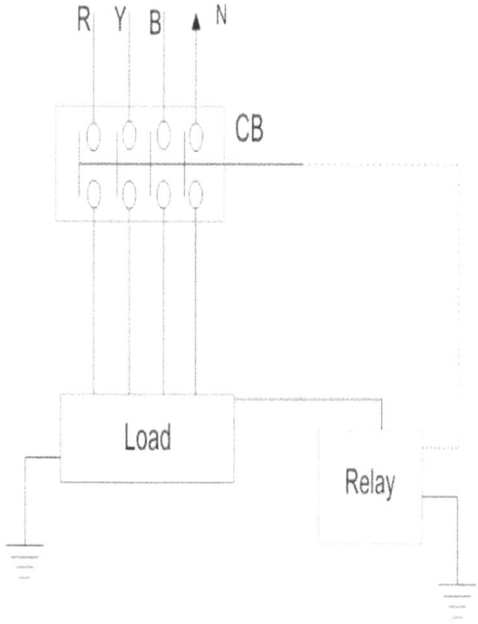

3

2- Current ELCB or RCD (Residual Current Device)

a- Type # A (two cores)

- This type of RCDs work in a similar way to electricity transformers
- Inside an RCD, the live and neutral cables from the electric supply wrap around an iron core much like the one in a transformer.
- The live cable wraps around one side of the core and the neutral cable goes around the other

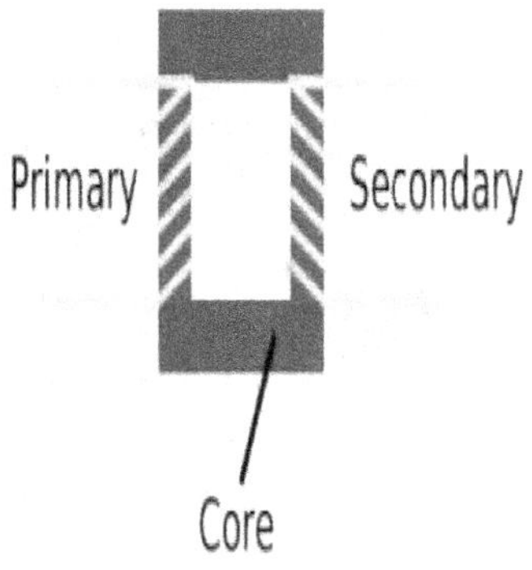

4

- Alternating current flows back and forth through the live wire (green).
- As it does so, it induces (creates) a magnetic field in the iron core, just like in a transformer (blue arrow).
- Meanwhile, an opposite alternating current is also flowing back and forth through the neutral wire (orange).

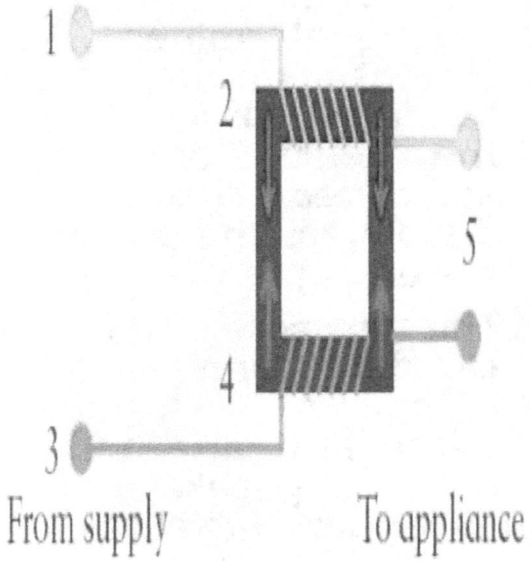

5

- The neutral current induces an equal and opposite magnetic field in the core (red arrow).
- Under normal conditions, the magnetic fields induced by the live and neutral wires cancel out: there is no overall magnetic field in the core and there's nothing to stop current flowing to the appliance you're using.

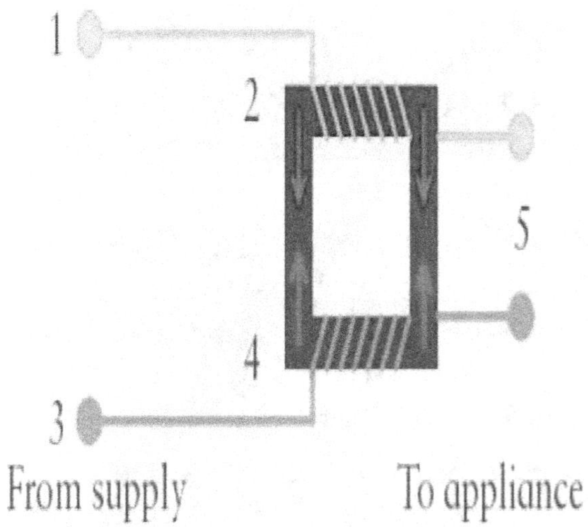

6

- The iron core has a third, smaller coil of wire wrapped around it.
- This is called the search or detector coil and it's wired up to a very fast electromagnetic switch called a relay.
- When a current imbalance occurs, the magnetic field induced in the core causes an electric current to flow in the search coil.
- That current triggers the relay, and the relay then cuts off the power.

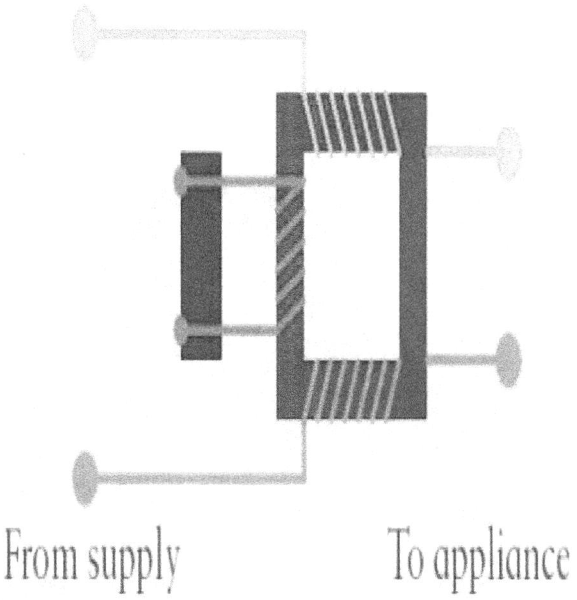

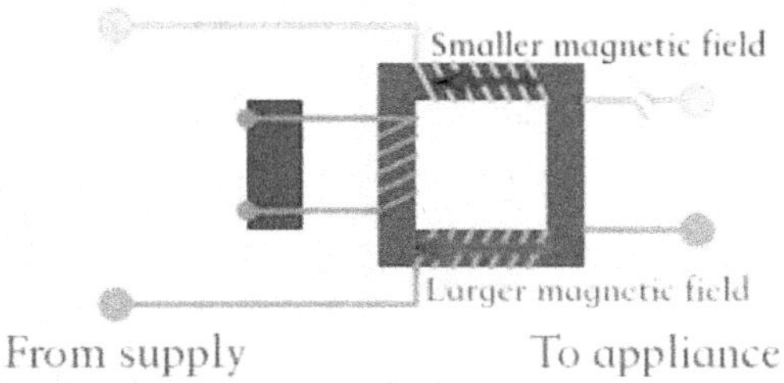

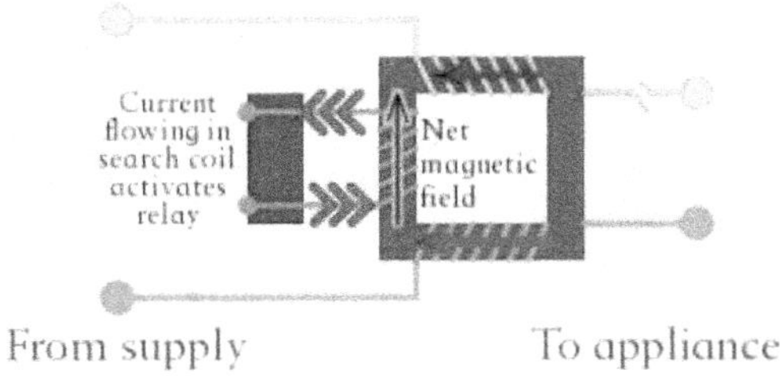

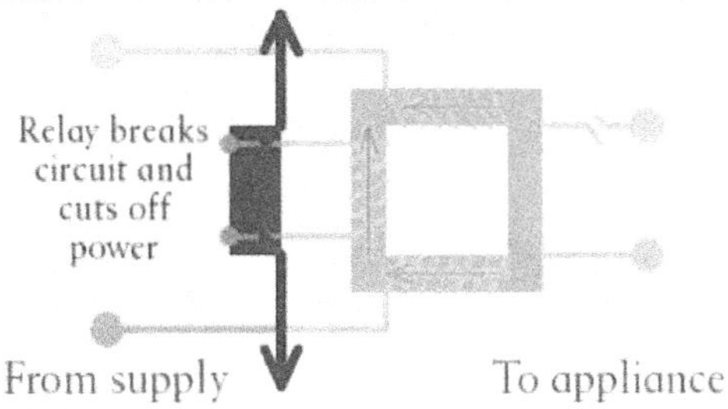

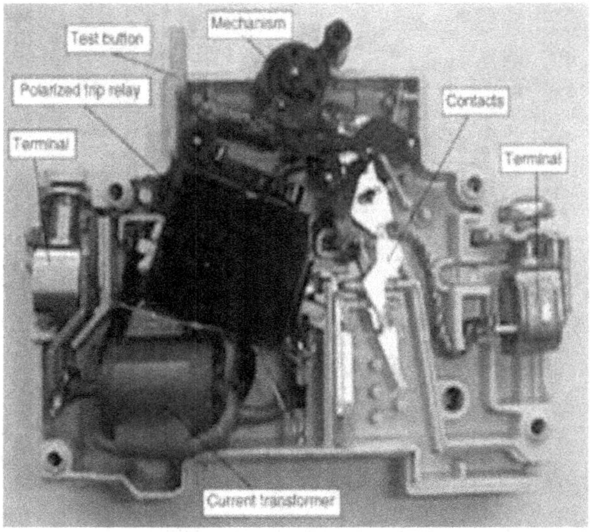

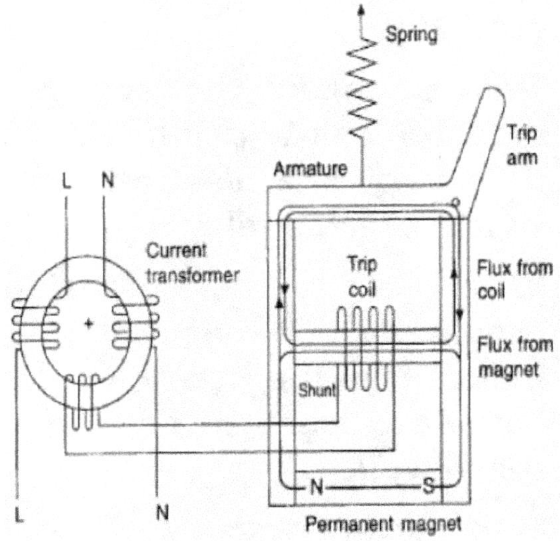

b-Type # B (one balanced core)

- This type of RCD devices measure the vector sum of currents passing through the phase and neutral conductors in a circuit, via a magnetic coil and electronic amplifier.
- The device will trip if these are out of balance by more than the residual operating current, in accordance with the manufacturer's time-current performance curve.
- RCD's contain a core-balance transformer

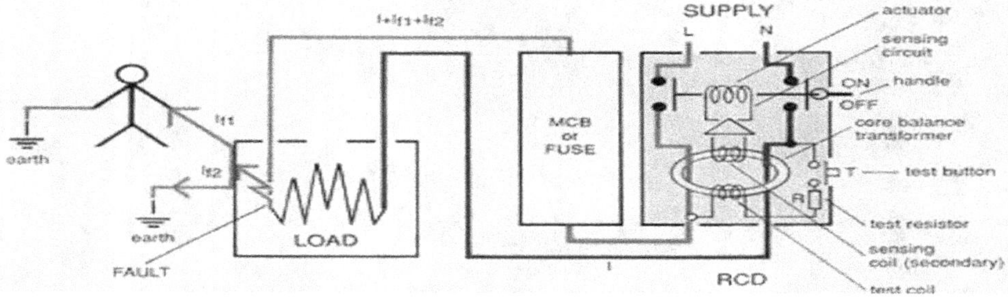

If the current which goes out on the phase does not return on the neutral, or on another phase, in the event of a three phase type, it will be detected as an earth fault through an imbalance in the core balance transformer

This imbalance will activate a tripping coil so that the circuit will be disconnected.

Therefore, if one comes in contact with a live appliance, the current, which flows through this person to return to earth, is sufficient to trip a 30mA RCD instantly.

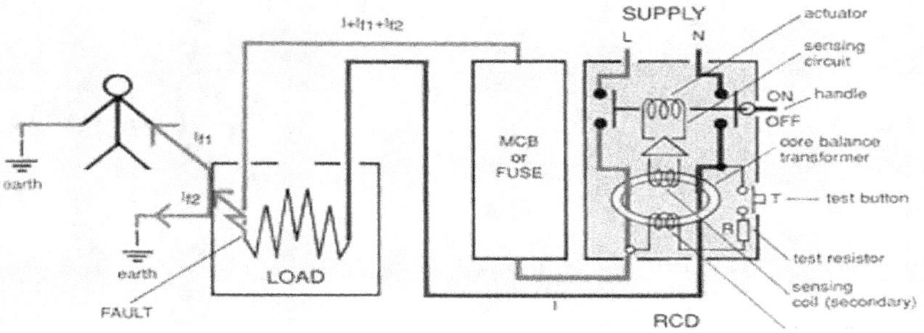

Three Phase Current ELCB or RCD (Residual Current Device)

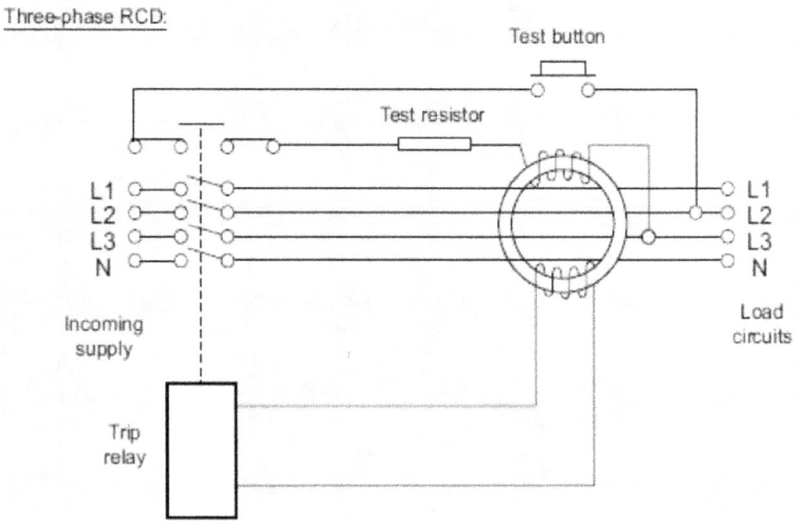

Single Phase Current ELCB or RCD (Residual Current Device)

Single-phase RCD:

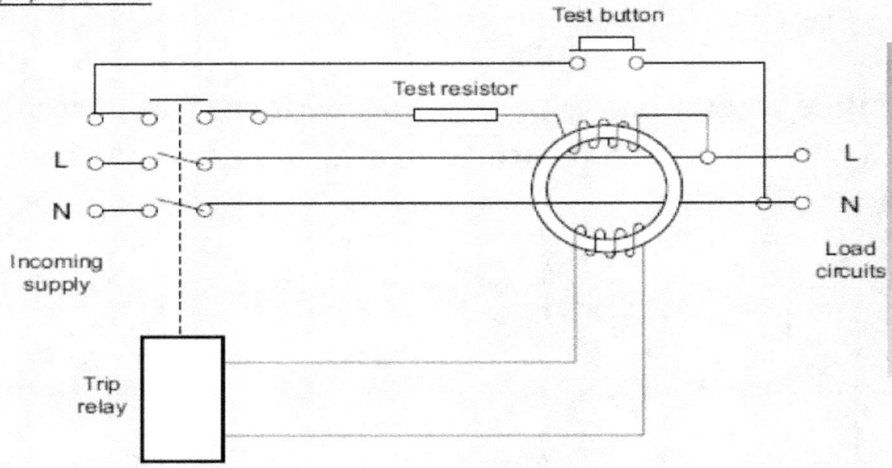

Type of fault	Reason
Downstream of RCD	• Direct Contact by persons • Indirect Contact (during earth fault) • Incorrect discrimination with upstream and downstream devices (e.g. between MDB and FDB) • Loose connections • Crossed neutral connections on split busbar distribution board • Neutral to earth fault • High Earth Conductor currents (e.g. IT equipment, filters, etc) • Moisture in Circuit conductors (especially joints in MICC cables) • Moisture in Appliances (e.g. cooker heating element) • Double pole switching (capacitive effects) • Transient voltages caused by large inductive loads (e.g. industrial motors) • Damaged Circuits (e.g. nails in walls)
Upstream of RCD	• Loose connections • Mains borne disturbances (e.g. surges, lightning, harmonics, transients from overhead lines) • Disturbing loads (e.g. machinery, lift motor, etc)

(7)

$P_s(3\phi) = 322.391$ MW

$Q_s(3\phi) = 288.609$ MVAR

$S_R(3\phi) = 3 V_R I_R^* = 3 \times 127 \times 10^3 \times 1000 \angle 36.87$

$= 381 \times 10^6 \angle 36.87$ VA $= 304.799 \times 10^6 + j\,228.6 \times 10^6$

$= P_R + j Q_R$

$\eta = \frac{P_R}{P_S} \times 100 = \frac{304.799 \times 10^6}{322.391 \times 10^6} \times 100 = 94.4\ \%$

Voltage regulation $= \frac{V_S - V_R}{V_R} \times 100 = \frac{144.33 - 127}{127} \times 100 = 13.646\ \%$

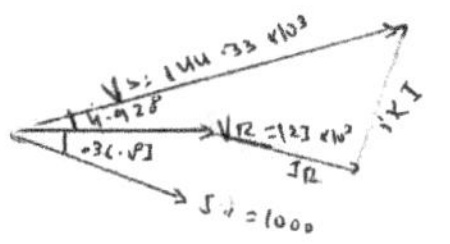

(b) Find the voltage at sending end and voltage regulation and efficiency when line supplies a 3 phase load 381 MVA of 0.8 PF leading.

$I_R = 1000 \angle 36.87$

$V_s = V_R + Z I$

$= 127 \times 10^3 + (6 + j20)\ 1000 \angle 36.87$

$= (119.8 + j\,19.6) \times 10^3$

$= 121.3927 \times 10^3 \angle 9.292$ V

$S_s = 3 V_s I_s^* = 3 \times 121.3927 \times 10^3 \angle 9.292 \times 1000 \angle -36.87$

$= 364.178 \times 10^6 \angle -27.578 = 322.8 \times 10^6 - j\,168.548 \times 10^6$

$= 364.178 \times 10^6 \angle -27.578$

$P_s(3\phi) =$

322.8×10^6 W

$= 322.8$ MW

Ex: (8)

A 220 kV (L-L) three phase line is 40 km long. The resistance per phase 0.15 Ω per km and inductance per phase is 1.3263 mH/km. Frequency = 60 Hz.

(a) Find the voltage and power at sending end and voltage regulation and efficiency when line supplies a 3 phase load 381 MVA of 0.8 PF lagging.

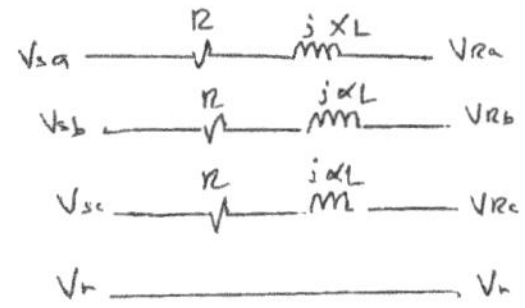

Solution:

$R = rl = 0.15 \times 40 = 6\ \Omega$

$L = 1.3263 \times 10^{-3} \times 40 = 5.3 \times 10^{-2}$

$X_L = 2\pi f L = 2\pi \times 60 \times 5.3 \times 10^{-2} = 20\ \Omega$

The receiving end voltage $V_{R(L-N)} = \frac{V_{R(L-L)}}{\sqrt{3}} = \frac{220\text{kV}}{\sqrt{3}} = 127.168 \times 10^3$

The apparent power $|S_R(3\phi)| = 381 \times 10^6$

$|S_R(3\phi)| = 3 |V_{R(L-N)}|\,|I_R|$

$|I_R| = \frac{381 \times 10^6}{3 \times 127 \times 10^3} = 1000$

$I_R = 1000 \angle -\cos^{-1} 0.8$

$= 1000 \angle -36.87$

$V_{s(L-N)} = V_{R(L-N)} + Z I$

$= 127 \times 10^3 + (6 + j20) \times 1000 \angle -36.87$

$= 144.333 \times 10^3 \angle 4.928$

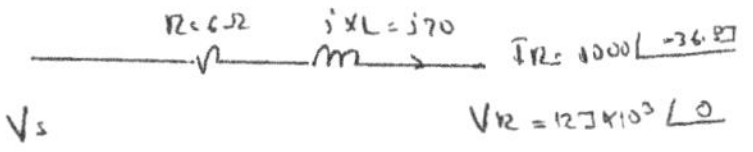

$S_s(3\phi) = 3 V_s I_s^* = 3 \times 144.33 \times 10^3 \angle 4.93 \times 1000 \angle 36.87$

$= 433 \angle 41.8 \times 10^6 = (322.391 + j\,288.609) \times 10^6$ VA

$= P_s + j Q_s$

Transformers

Dr. Hidaia alassouli

Transformer

- The ideal transformer
- Vp ~=>ip ~ => Ø~=> induced voltage in primary
- => induced voltage in secondary winding
- The induced voltage in the primary winding
- $v_p = N_p \frac{d\emptyset}{dt}$
- The induced voltage in the secondary winding
- $v_s = N_s \frac{d\emptyset}{dt}$
- $\frac{\text{Vp}}{\text{Vs}} = \frac{Np}{Ns} = a$

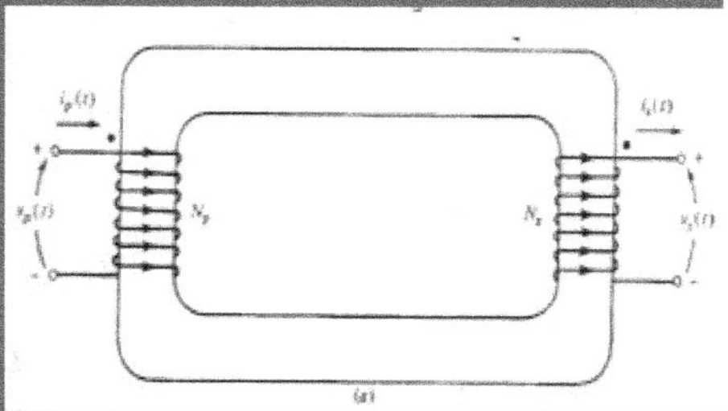

- The current equation
- $N_p i_p - N_s i_s = \emptyset R$
- $R = \frac{lc}{\mu A}$
- Reluctance of core very small =R=0
- $N_p i_p - N_s i_s = 0$
- $\frac{ip}{is} = 1/a$

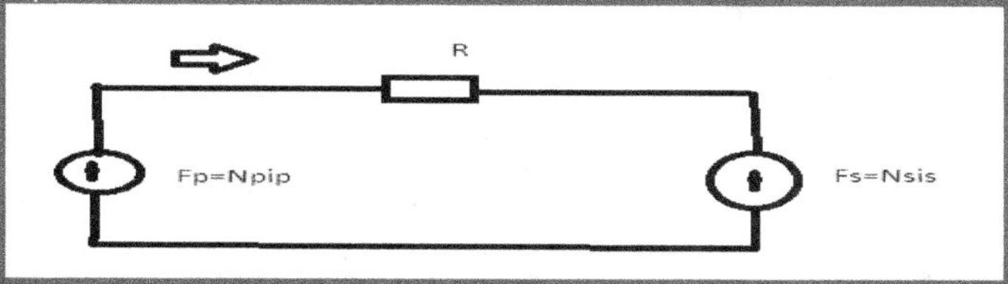

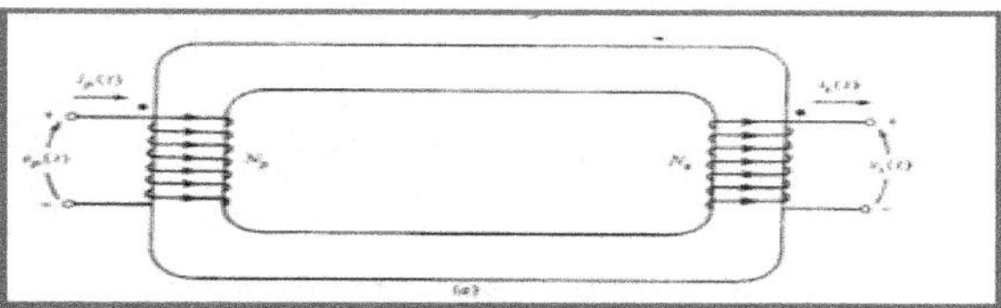

- The active power equation:

$$P_{\text{in}} = V_P I_P \cos \theta_P$$

$$P_{\text{out}} = V_S I_S \cos \theta_S$$

$$P_{\text{out}} = \frac{V_P}{a} a I_P \cos \theta$$

$$P_{\text{out}} = V_P I_P \cos \theta = P_{\text{in}}$$

- The reactive power equation

$$Q_{\text{in}} = V_P I_P \sin \theta = V_S I_S \sin \theta = Q_{\text{out}}$$

$$S_{\text{in}} = V_P I_P = V_S I_S = S_{\text{out}}$$

Referring the load to primary

- Referring the load from the secondary to primary

$$Z_L = \frac{\mathbf{V}_S}{\mathbf{I}_S}$$

$$V_P = aV_S$$

$$\mathbf{I}_P = \frac{\mathbf{I}_S}{a}$$

$$Z'_L = \frac{\mathbf{V}_P}{\mathbf{I}_P}$$

$$Z'_L = \frac{\mathbf{V}_P}{\mathbf{I}_P} = \frac{a\mathbf{V}_S}{\mathbf{I}_S/a} = a^2\frac{\mathbf{V}_S}{\mathbf{I}_S}$$

$$Z'_L = a^2 Z_L$$

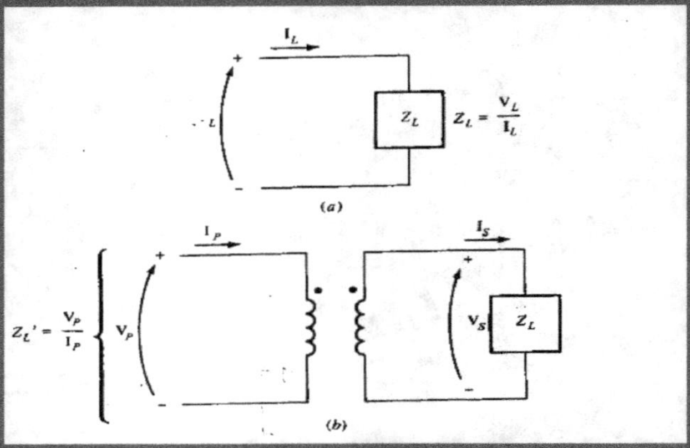

Equivalent Circuit of Non Ideal Transformer

- The non ideal transformer equivalent circuit

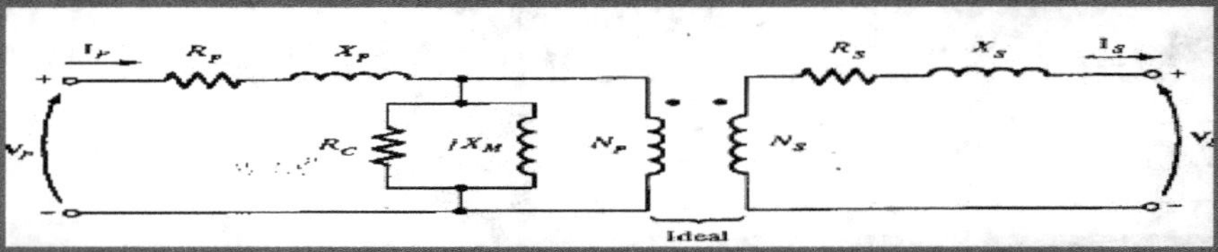

- The equivalent circuit referred to primary

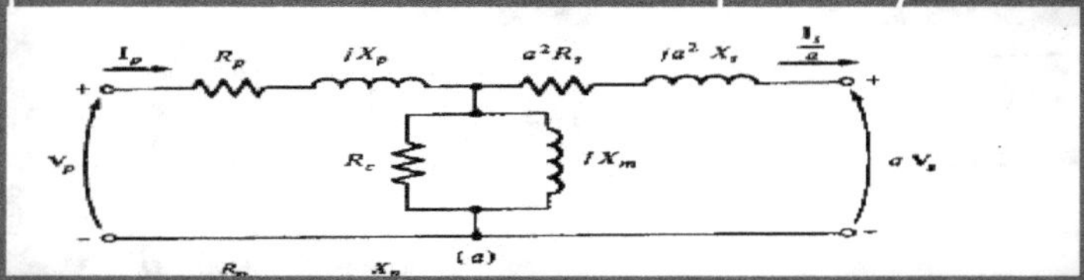

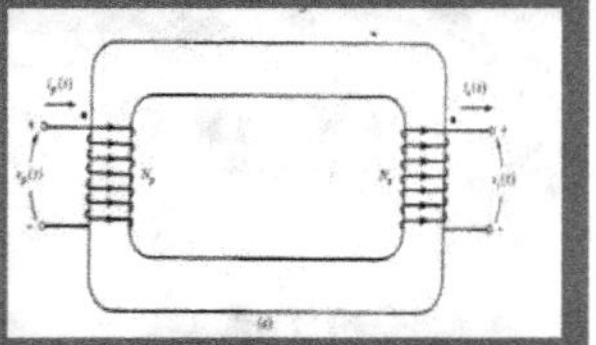

- The equivalent circuit referred to secondary

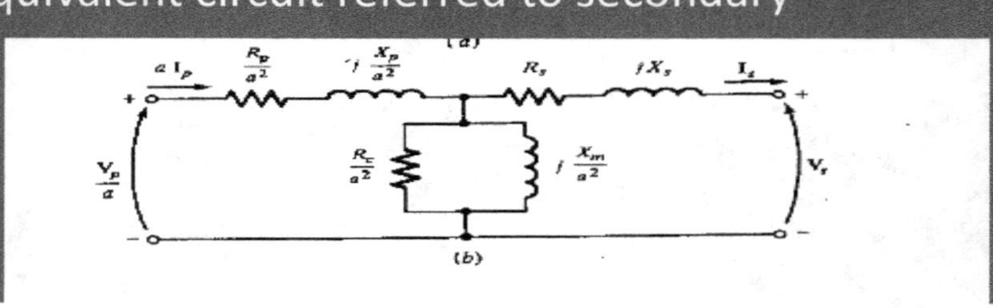

Three phase transformer

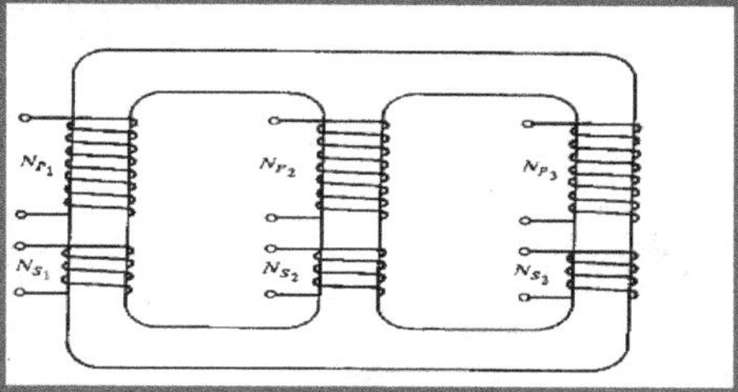

- Wye-Wye Connection

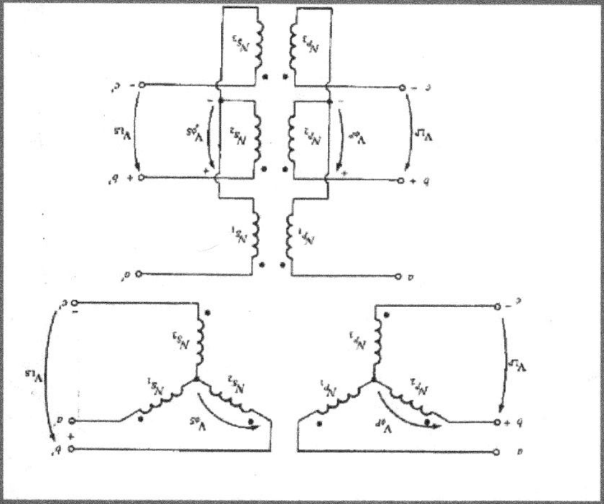

$$\frac{V_{LP}}{V_{LS}} = \frac{\sqrt{3}V_{\phi P}}{\sqrt{3}V_{\phi S}} = a$$

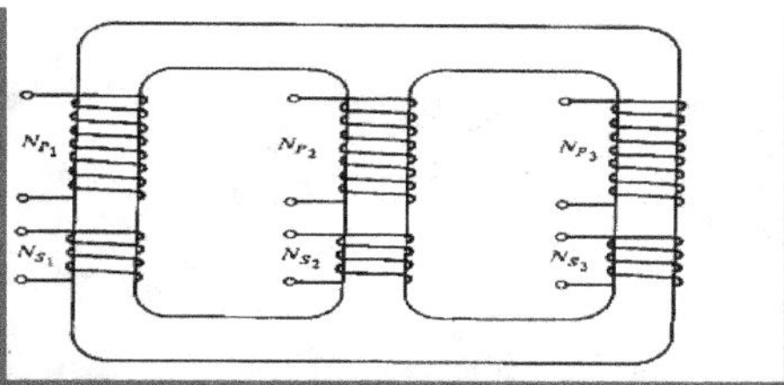

- Wye Delta Connection

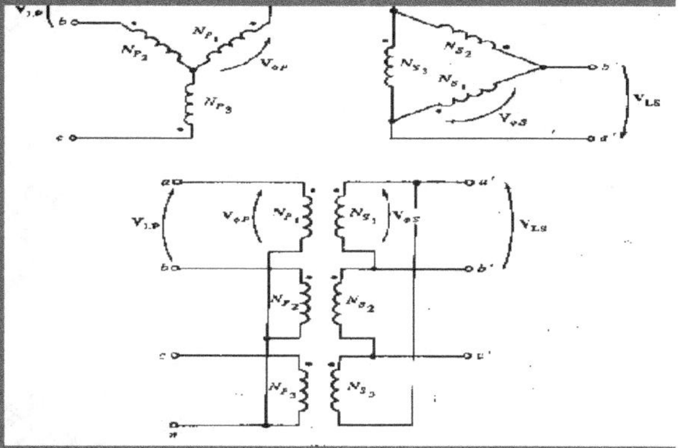

$$\frac{V_{LP}}{V_{LS}} = \frac{\sqrt{3}V_{\phi P}}{V_{\phi S}}$$

$$\frac{V_{LP}}{V_{LS}} = \sqrt{3}a \qquad \text{Y-}\Delta$$

- Delta Wye Connection

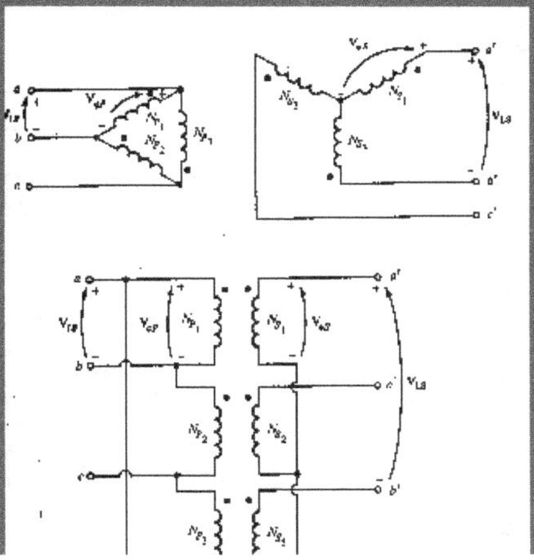

$$\frac{V_{LP}}{V_{LS}} = \frac{V_{\phi P}}{\sqrt{3}V_{\phi S}}$$

$$\frac{V_{LP}}{V_{LS}} = \frac{a}{\sqrt{3}} \qquad \Delta\text{-Y}$$

• Delta Delta Connection

$$V_{LP} = V_{\phi P}$$
$$V_{LS} = V_{\phi S}$$

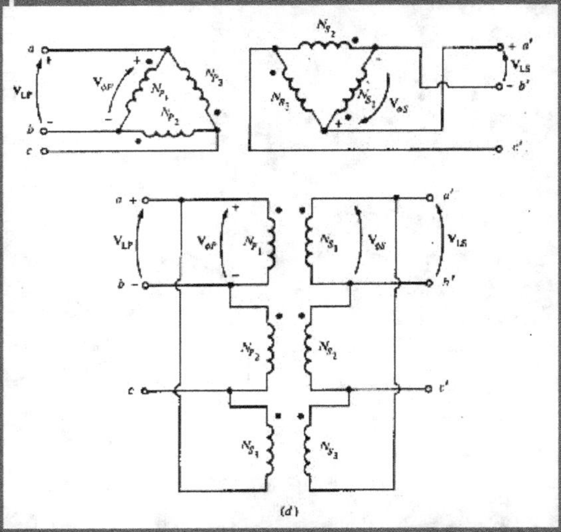

www.ingramcontent.com/pod-product-compliance
Lightning Source LLC
Chambersburg PA
CBHW081558220526
45468CB00010B/2690
9781984050335